Robert H Cousins

A Theoretical and Practical Treatise on the Strength of Beams and

Columns

Robert H. Cousins

A Theoretical and Practical Treatise on the Strength of Beams and Columns

ISBN/EAN: 9783337106409

Printed in Europe, USA, Canada, Australia, Japan

Cover: Foto ©berggeist007 / pixelio.de

More available books at **www.hansebooks.com**

A Theoretical and Practical Treatise

ON THE

STRENGTH

OF

BEAMS AND COLUMNS;

IN WHICH

THE ULTIMATE AND THE ELAS-
TIC LIMIT STRENGTH OF BEAMS AND COL-
UMNS IS COMPUTED FROM THE ULTIMATE AND ELASTIC
LIMIT COMPRESSIVE AND TENSILE STRENGTH OF THE MATERIAL,
BY MEANS OF FORMULAS DEDUCED FROM THE CORRECT
AND NEW THEORY OF THE TRANSVERSE
STRENGTH OF MATERIALS.

BY

ROBERT H. COUSINS,

Civil Engineer,

Formerly Assistant Professor of Mathematics at the Virginia Military Institute,
Lexington, Va.

E. & F. N. SPON,

NEW YORK :   LONDON :
12 CORTLANDT STREET.   125 STRAND.
1889.

INTRODUCTION.

For more than two centuries the mathematical and mechanical laws that govern the transverse strength of Beams and Columns have received the attention of the most expert mathematicians of all countries. Galileo, in 1638, formulated and published the first theory on the subject. He was followed by such philosophers as Mariotte, Leibnitz, Bernouilli, Coulomb, and others, each amending and extending the work of his predecessor, until the year 1824, when Navier succinctly stated the theory that is recognized to be correct at the present day, and to which subsequent writers and investigators have added but little.

This theory has neither received the endorsement of the experimenters nor of some of the theoretical writers. "Excepting as exhibiting approximately the laws of the phenomena, the theory of the strength of materials has many practical defects" (Wiesbach). "It has long been known that under the existing theory of beams, which recognizes only two elements of strength—namely, the resistance to direct compression and extension—the strength of a bar of iron subjected to a transverse strain cannot be reconciled with the results obtained from experiments on direct tension, if the neutral axis is in the centre of the bar" (Barlow).

During the present century much time and means have been expended in attempts to solve, experimentally, the problems that have engaged the attention of the mathematicians, and as the result of their labors we find such experimenters as Hodgkinson, Fairbairn, and others, whose names are household words in the literature of the subject, adopting *empirical* rules for the strength of Beams and Columns rather than the *rational* formulas deduced by the scientists. "For no theory of the rupture of a simple beam has yet been proposed which fully satisfies the critical experimenter" (De Volson Wood).

That we should be able to deduce the strength of Beams and Columns from the known *tensile* and *compressive* strength of the material composing them, has been apparent to many writers and experimenters on the subject, but to the present time no theory has been advanced that embodies the mathematical and mechanical principles necessary to its accomplishment. The theory herein advanced and the formulas resulting therefrom deduce the strength of Beams and Columns from the direct *crushing* and *tensile* strength of the material composing them, without the aid of that coefficient that has no place in nature, *the Modulus of Rupture*. The theory and the formulas deduced therefrom are in strict accord with correct mechanical and mathematical principles, and the writer believes that they will fully satisfy the results obtained by the experimenter.

The great practical benefits to be derived from the correct theory of the strength of Beams and Columns will be evident, when we consider the countless tons of metal that have been made into railroad rails, rolled beams, and the other various shapes, and that the manufacturers were without knowledge of the work to be performed by the different parts of the beam or column in sustaining the load that it was intended to carry. The best that they have been able to do is to compute the strength by the aid of an *empirical* quantity deduced from experiments on "similar beams." The correct theory will enable them to foretell the strength of any untried shape, and the reason for the strength of those that have been long in use, which is the "true object of theory."

R. H. C.

DALLAS, TEXAS, *March* 13, 1889.

CONTENTS.

CHAPTER I.

FORCES DEFINED AND CLASSED.

CONTENTS.

CHAPTER III.

TRANSVERSE STRENGTH.

CHAPTER IV.

CAST-IRON BEAMS.

CONTENTS. vii

Note.—The writer regrets that chapters on the following subjects are unavoidably omitted at this time, "The Strength of Arches" and "The Deflection of Beams," as they were not satisfactorily complete, and "Beams of Maximum Strength with Minimum Material," as he was not fully protected by letters-patent at the time of going to press.

STRENGTH OF BEAMS AND COLUMNS.

CHAPTER I.

FORCES.

Section I.—*Forces Defined and Classed.*

1. Force defined. Force may be defined to be an action between two bodies, or parts of a body, which either causes or tends to cause a change in their condition, whether of rest or of motion; and our knowledge of a *force* is practically complete when we know its *intensity*, expressed in pounds or some other unit of measure, its *direction*, whether toward or from the body upon which it acts, its *point* of *application*, and the *angle* that its *line* of *direction* makes with the surface of the body.

For an *unit* of *measure* of forces we shall use the pound avoirdupois, as all of our recorded knowledge of the strength of materials used in structures is expressed in *pounds on the square inch.*

2. Stress and **Strain** are words used to define that class of forces that are brought into action when contiguous parts of a body are caused to *react* upon each other by reason of the application of other forces to the body, and may be classed as follows:

Compression, Thrust, or *Pressure* is the force *exerted* when the contiguous parts of a body are caused to move *toward* each other.

Tension, Pull, or *Tensile Strain* is the force *exerted* when the contiguous parts of a body are caused to move *away* from each other.

3. The Load. The external forces applied to a body to produce the various kinds of *stress* or *strain* is called the *Load,* and its amount or magnitude is expressed in pounds ; it may be classed as follows :

Concentrated Forces or *Loads.* While in nature every force must be distributed over a definite amount of surface, it is necessary, in order to define certain principles, to consider them to be *concentrated* at a single point, the effect being identical with that of the distributed load.

A *Distributed Force* of *uniform intensity* is the force or load that acts with the same intensity on each square inch of the surface of the body over which it is distributed.

A *Distributed Force* of *uniformly varying intensity* is a force that increases in intensity in direct proportion to the distance from a given point.

4. Equilibrium and **Resultant.** *Equilibrium* of a *system* of *forces* is such a condition that the combined action of the forces produces no change in the rest or *motion* of the body to which it is applied.

The *Resultant* of a *system* of *forces* is a single *concentrated* force that will produce the same effect upon the body, if applied, that the system of forces will produce.

SECTION II.—*Concentrated Parallel Forces.*

5. Bending Moment. In this section will be deduced the relation existing between the *vertical* forces caused by *gravity*, the supporting forces and the horizontal distances between their lines of action. In order to ascertain the relation between such forces it is necessary to obtain the product of each force by the perpendicular distance of its line of action from a given point. Such product is called the *Bending Moment* of the force.

The *vertical* forces will be called the *loads*, and the connection between their lines of action will be called a *beam*, without reference to the *shape* of its cross-section, its material, or its ability to resist the *bending moment* of the applied loads, which will be considered in the sequel.

The following *notation* will be used throughout :

L = the *total* applied load in pounds.

s = the *span*, the horizontal distance between the supports in inches.

M = the *bending moment.*　　　　　　　　•

Case I.—THE BEAM FIXED AT ONE END AND LOADED AT THE OTHER.

A load thus applied will *bend* the beam (Fig. 1) at each

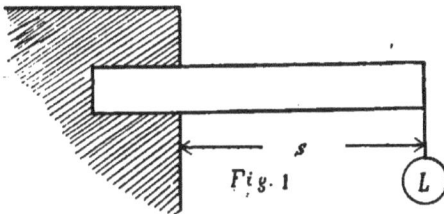

Fig. 1

section from the free end to the point of support, but unequally. The bending moment at any section is the product of·

the load by the distance of the section from the free end of
the beam. The Greatest Bending Moment, at the point of
support, will be

$$GBM = L \times s.$$ (1)

Case II.—THE BEAM FIXED AT ONE END AND THE LOAD
UNIFORMLY DISTRIBUTED OVER THE ENTIRE SPAN.

The *bending moment* at *any* section is equal to the product
of the load between the free end of the beam and the section
by *one half* of its distance from the free end; the Greatest
Bending Moment, at the point of support, will be

$$GBM = L \times \frac{s}{2},$$ (2)

or *one half* what it is in Case I., the load, *L*, and the span, *s*,
being the same.

Case III.—THE BEAM SUPPORTED AT THE ENDS AND THE
LOAD APPLIED AT THE MIDDLE OF THE SPAN.

In order that *equilibrium* shall exist one half of the load
must be supported by each point of support; the load will

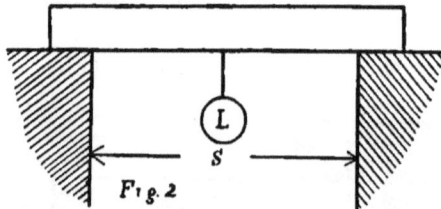

Fig. 2

bend the beam (Fig. 2) in the same manner that it would if
the beam were *fixed* in the middle of its span and loaded at
each free end with *one half* of the applied load, *L*; the *bending moment* at *any* section will be one half of the load
multiplied by its distance from the nearest point of support;

the Greatest Bending Moment, at the middle of the span, will be

$$GBM = \frac{L}{2} \times \frac{s}{2} = \frac{L \times s}{4}, \qquad (3)$$

or *one fourth* of that in Case I., and *one half* that in Case II., the load and span being the same in each.

Case IV.—THE BEAM SUPPORTED AT THE ENDS AND THE LOAD, L, UNIFORMLY DISTRIBUTED OVER THE SPAN.

The Greatest Bending Moment occurs at the middle of the span,

$$GBM = \frac{L \times s}{8}. \qquad (4)$$

Case V.—THE BEAM FIXED AT BOTH ENDS AND THE LOAD APPLIED AT THE MIDDLE OF THE SPAN.

In the preceding cases the *bending moment* of the applied load, L, produces a strain of compression in either the top or the bottom of the beam and a tensile strain in the opposite side, but in this and the next case the bending moment pro-

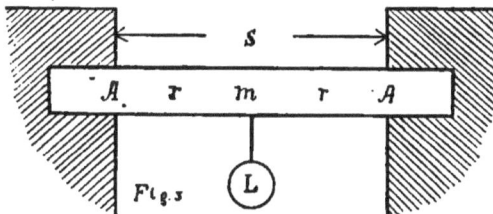

duces in the upper side, over each point of support, AA (Fig. 3), a tensile strain, and in the middle a compressive strain; and in the lower side at each point of support, AA, a compressive strain, and in the middle, m, a tensile strain. Now, in order that these directly opposite strains may exist in the upper and lower sides of the beam at the same time, the strains at one section on each side of the middle section must be zero in intensity. At these points, *rr*, the *curvature* of the beam changes; they may be called the points of *reverse curvature*, and in this case are located at *one fourth* of the span from each point of support, *AA*.

The *Bending Moments* are equal and greatest at three sections; at each point of support and at the middle of the span, *theoretically* its value is given by the following equation:

$$GBM = \frac{L \times s}{8}. \tag{5}$$

Barlow and other experimenters state that this should be

$$GBM = \frac{L \times s}{6}.$$

Case VI.—THE BEAM FIXED AT BOTH ENDS AND THE LOAD DISTRIBUTED UNIFORMLY OVER THE SPAN.

The sections at which the Greatest Bending Moment occurs are the same as in the preceding case; but the points of reverse curvature, *rr*, are at the distance 0.2113*s* from each point of support, *AA*, Fig. 3.

At the middle of the span, *m*,

$$GBM = \frac{L \times s}{24}. \tag{6}$$

At the points of support, *AA*,

$$GBM = \frac{L \times s}{12}. \tag{7}$$

6. Case in General. Theory has demonstrated and experiments fully confirm, except as previously noted, that the following relations exist between the Greatest Bending Moments in beams, the span and the total applied load being the same—that in a beam fixed at one end and loaded at the other being taken as our *unit* or *standard* of measure :

n

Beam fixed at one end and loaded at the other............ 1

Beam " " " " "   "   uniformly.............. $\frac{1}{2}$

Beam supported at the ends and loaded at the middle...... $\frac{1}{4}$

Beam   "   " " " "   "   uniformly......... $\frac{1}{8}$

Beam fixed at both ends and loaded at the middle......... $\frac{1}{8}$

Beam " " " " "   "   uniformly ....$\frac{1}{12}$

From which we deduce the following formula, applicable to all of the preceding cases :

$$GBM = L \times s \times n. \qquad (8)$$

Placing for the factor, n, the values given in the preceding table of comparison, the formulas heretofore deduced for each case will be reproduced.

These formulas for the Greatest Bending Moments are entirely independent of the material composing the beam and of its cross-section.

SECTION III.—*Uniformly Varying Forces—Rectangular Areas.*

7. Notation. The principles governing the action of uniformly varying forces distributed over rectangular surfaces will be deduced, and then those applicable to surfaces bounded by curved lines will be considered. In order to determine the effect of such a force, we must determine the resultant, its point of application, its lever-arm, and from these the moment of the uniformly varying force. The following notation will be used, and it will have the same meaning whenever it appears in the following pages :

$C =$ the maximum compressive strain, in pounds, per square inch.

$T =$ the maximum tensile strain, in pounds, per square inch.

$R_x =$ the moment of the tensile strain in inch-pounds.

$R_c =$ the " " " compressive strain in inch-pounds.

$d_c =$ the depth of the area covered by the compressive strain in inches.

$d_x =$ the depth of the area covered by the tensile strain in inches.

$d = d_c + d_x =$ the total depth of the area in inches.

$b =$ the total width of the area in inches.

8. Resultant. The amount of *direct* strain is equal to

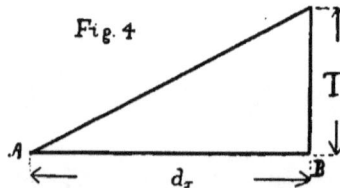

Fig. 4

the weight of a prismoidal wedge composed of such material that a prism an inch square in section and C or T high will give a pressure on its base equal to C or T as defined above; the *resultant* will then be equal to the product of the depth d_c or d_r, by the width b, by *one half* of the height C or T, which is the volume of the wedge, and it will pass through a point at $\frac{2}{3}$ the depth d_c or d_r from the edge of the wedge A, of which Fig. 4 is a section.

$$\therefore \ Resultant = \frac{bd_r T}{2} \ or \ \frac{bd_c C}{2}. \qquad (9)$$

$$Lever\text{-}arm = \frac{2d_r}{3} \ or \ \frac{2d_c}{3}. \qquad (10)$$

By the Calculus:

· Let $x_r = d_r = AB$, the depth of the area,

$x = $ any distance from A,

$\dfrac{Tx}{x_r} = $ the height of the wedge at x, distance from A,

$bdx = $ a small area, or the differential,

$$Volume = \int_0^{x_r} \frac{Tx}{x_r} \cdot bdx = \frac{Tbx^2}{2x_r},$$

$$\therefore \ Resultant = \frac{Tbd_r}{2},$$

$$The \ moment = \int_0^{x_r} \frac{Tx^2}{x_r} \cdot bdx = \frac{Tbx^3}{3x_r},$$

$$\therefore \ The \ moment = \frac{Tbd^2_r}{3}.$$

Dividing the moment by the resultant we obtain the lever arm $= \frac{2}{3}d_r$.

9. Problem I. REQUIRED THE MOMENT OF AN UNIFORMLY VARYING COMPRESSIVE FORCE OF MAXIMUM INTENSITY, C, WITH RESPECT TO AN AXIS, B, IN THE BACK OF THE PRESSURE WEDGE.

Let ABD (Fig. 5) represent a section through the pressure wedge.

$$The \ resultant = \frac{bd_c C}{2},$$

The lever-arm $= \frac{1}{3}d_c$,

$$\therefore R_c = \frac{bd_c^2 C}{6}. \tag{11}$$

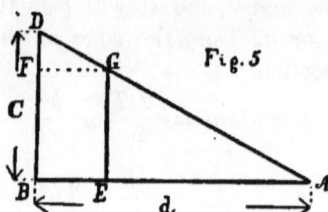

Fig. 5

PROBLEM II. *Required the moment of only a portion of the force distributed over a depth $EB = d_1$ and width b_1,— the force being zero in intensity at A; the axis being at B* (Fig. 5).

$\dfrac{d_c - d_1}{d_c} \cdot C \doteqdot$ the intensity of the force at E,

$\dfrac{d_1}{2} =$ the lever-arm of this force at E, considered

to be constant over the distance EB,

$\left(1 - \dfrac{d_c - d_1}{d_c}\right) C =$ the intensity of the force at B less that

at E,

$\dfrac{d_1}{3} =$ its lever-arm.

Having considered the force to be divided into two portions, the first, GE, constant in intensity over the depth EB, and the second increasing in intensity from zero at G to $\left(1 - \dfrac{d_c - d_1}{d_c}\right) C$ at F, by adding the moments of these *parts* we obtain

$$R_c = \frac{b_1 d_1^2}{6 d_c} (3d_c - 2d_1)C. \tag{12}$$

By the Calculus :

Let $x_c = d_c =$ the depth AB (Fig. 5),

$x =$ any distance from A,

$\dfrac{x_c - x}{x_c} \cdot C =$ the intensity of the force at the depth x,

$b dx =$ a small area of the base of the wedge,

$$\therefore R_c = \int_0^{x_c} \frac{x\,(x_c - x)}{x_c} \cdot C \cdot b dx = b\left(\frac{x_c\,x^2}{2x_c} - \frac{x^3}{3x_c}\right) C,$$

$$\therefore R_c = \frac{b d_c^2 C}{6}, \text{ or Eq. 11.}$$

Integrating the above expression for R_c, between the limits $x = d_c$ and $x = d_1$, we obtain

$$R_c = \frac{b_1 d_1^2}{6\,d_c}\,(3d_c - 2d_1)\,C, \text{ or Eq. 12.}$$

10. Problem III. REQUIRED THE MOMENT, R_τ, OF AN UNIFORMLY VARYING TENSILE FORCE, WITH RESPECT TO AN AXIS, O, PARALLEL TO THE EDGE, A, OF THE TENSION WEDGE, AND AT THE DISTANCE $OA = d_c$ FROM IT.

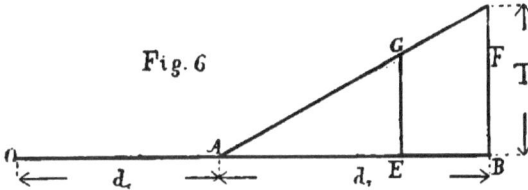

Fig. 6

Let ABD (Fig. 6) represent a section through the tension wedge.

$$\frac{b d_\tau T}{2} = \text{the resultant,}$$

$$d_c + \tfrac{2}{3} d_\tau = \text{its lever-arm,}$$

$$\therefore R_\tau = (d_c + \tfrac{2}{3} d_\tau)\,\frac{b d_\tau T}{2},$$

or,

$$R_\tau = \frac{b d_\tau}{6}\,(2d + d_c)\,T. \tag{13}$$

PROBLEM IV. *Required the moment of only a part of the above force distributed over the depth* $EB = d_2$ *and the width* b_2.

Consider the force to be divided into two portions. The first, with the intensity at E constant over the distance EB; the second increasing in intensity from zero at G to $\dfrac{d_2}{d_\tau} T$ at F.

$$d_2 \left(1 - \frac{d_2}{d_\tau}\right) T = \text{the intensity of the force at } E,$$

$$\left(d - \frac{d_2}{2}\right) = \text{its lever-arm},$$

$$\frac{d_2}{d_\tau} T = \text{the intensity of the force at } B, \text{ less that } E,$$

$$\left(d - \frac{d_2}{3}\right) = \text{its lever-arm}.$$

By adding the moments we obtain

$$R_\tau = \frac{b_2}{6d_\tau}\left[3d_2 d_\tau (2d - d_2) - d_2^2(3d - 2d_2) \right]. \qquad (14)$$

By the Calculus :

Let $x_\tau = d_\tau = $ the distance AB,

$x = $ any distance from A,

$\dfrac{xT}{x_\tau} = $ the intensity of the force at any distance x,

$(d_c + x) = $ its lever-arm,

$bdx = $ the differential of the area of the base,

$$\therefore R_\tau = \int_{x_\tau}^{0} \frac{xT}{x_\tau} (d_c + x)\, bdx = \frac{b}{x_\tau}\left(\frac{d_c x^2}{2} + \frac{x^3}{3}\right) T,$$

$$\therefore R_\tau = \frac{bd_\tau}{6}(2d + d_c)\, T, \text{ or Eq. 13.}$$

Integrating the above expression for R_τ, between the limits $x = d_\tau - d_2$ and $x = d_\tau$, and placing b_2 for b, we have

$$R_\tau = \frac{b_\eta}{6d_\tau}\left[\,6d_\tau d_\vartheta(d-d_\vartheta) - d_\vartheta{}^2(3d_\varsigma - 2d_\vartheta).\,\right]T.$$

This is not identical in form with Eq. 14, but gives the same numerical result.

SECTION IV.— *Uniformly Varying Forces—Circular Segment Areas.*

11. Resultant and **Lever-arm.** The resultant and the moment of an uniformly varying force, distributed over circular segment areas, cannot be readily deduced by means of the elementary methods used when the areas were rectangular; but this can be accomplished with the aid of the Calculus.

12. Problem. REQUIRED THE MOMENT OF AN UNIFORMLY VARYING FORCE DISTRIBUTED OVER A CIRCULAR SEGMENT, WITH RESPECT TO AN AXIS THAT IS A TANGENT TO THE BASE OF THE PRESSURE WEDGE AND PARALLEL TO ITS EDGE, WHICH IS A CHORD OF THE CIRCLE.

Let, in Fig. 7, OBS represent the axis, MNB the base of the pressure wedge, and ABD a section through AB.

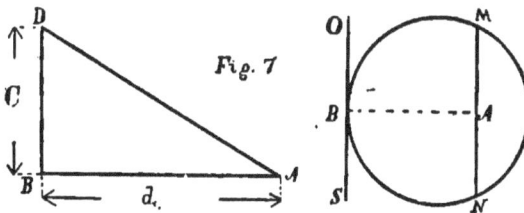

Fig. 7

Notation.—Adopting the notation defined in Art. 7 with the following in addition:

Let $r =$ the radius of the circle,

$d = d_o + d_\tau =$ the diameter,

$x_c = d_c =$ the versine of one half the arc MBN,

$x =$ any distance from the axis OBS,

$\dfrac{x_c - x}{x_c} \cdot C =$ the intensity of the pressure at any distance,

x, from the axis OBS,

$2 (2rx - x^2)^{\frac{1}{2}}dx =$ the differential of the area of the base,

$$\therefore R_c = \int_0^{x_c} \left(\frac{x_c - x}{x_c}\right) Cx \times 2 (2rx - x^2)^{\frac{1}{2}}dx.$$

Integrating, we obtain

$$R_c = 2C \left[(2rx_c - x^2_c)^{\frac{1}{2}} \left(\frac{x^2_c}{12} - \frac{rx_c}{12} - \frac{7r^2}{24} + \frac{15r^2}{24r_c}\right) + r\ versin.\ \frac{^{-1}x_c}{r} \left(\frac{r^2}{2} - \frac{15r^3}{24x_c}\right) \right].$$

Substituting the following equalities:

$$(2rx_c - x^2_c)^{\frac{1}{2}} = \sqrt{d_c d_\tau},$$

$r\ versin.\ \dfrac{^{-1}x_c}{r} =$ one half of the arc of the segment MBN

$$x_c = d_c\ ;$$

reducing, we obtain

$$R_c = \frac{C}{24d_c} \left[\sqrt{d_c d_\tau} \left[4d^2_c(d_c - r) + r^2(30r - 14d_c) \right] + Comp.arc\ (12d_c - 15r)r^2 \right] (15)$$

from which the moment of any compressed wedge may be computed. The factor Comp. Arc is the arc MBN.

13. Problem. Required the moment of an uniformly varying tensile force, distributed over the segment of a circle, with respect to an axis that is a tangent to the circle, parallel to the edge of the tension wedge, and at the distance d_c from it.

Let, in Fig. 8, FOS represent the axis, MNB the base of the tension wedge, MN the edge, and $OABD$ a section through OB.

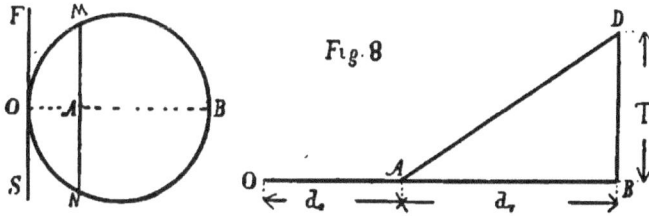

Fig. 8

Let $x_\tau = d_\tau =$ the distance AB,

$x =$ any distance from B toward A.

$\dfrac{x_\tau - x}{x_\tau} . T =$ the intensity of the force at any distance $x_\tau - x$

from A,

$d - x =$ its lever-arm.

$2 (2rx - x^2)^{\frac{1}{2}} dx =$ the differential of the area of the base of the wedge.

For notation, refer to Arts. 7 and 12.

$$R_\tau = \int_0^{x_\tau} \frac{(x_\tau - x)}{x_\tau} T (d - x) 2 (2rx - x^2)^{\frac{1}{2}} dx,$$

$$\therefore R_\tau = 2T\left[(2rx_\tau - x^2_\tau)^{\frac{1}{2}} \left(\frac{15rx_\tau}{12} - \frac{9r^2}{24} - \frac{2x^2_\tau}{24} + \frac{9r^3}{24x_\tau} \right) \right.$$

$$\left. + r \, versin.^{-1} \frac{x_\tau}{r} \left(\frac{3r^2}{6} - \frac{9r^3}{24x_\tau} \right) \right].$$

Substituting the following values:

$$(2rx_\tau - x^2_\tau)^{\frac{1}{2}} = \sqrt{d_c d_\tau},$$

$r \, versin.^{-1} \dfrac{x_\tau}{r} =$ one half the tension arc MBN,

$x_\tau = d_\tau$;

reducing, we obtain

$$R_\tau = \frac{T}{24d_\tau} \left[\sqrt{d_c d_\tau} \left[4d^2_\tau (5r - d_\tau) + 18r^2 (r - d_\tau) \right] \right.$$

$$\left. + \textit{Tension arc } (12d_\tau - 9r)r^2) \right] \ (16)$$

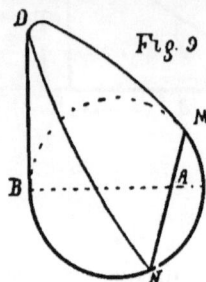

Fig. 9

from which the required moment may be computed.

14. Problem. REQUIRED THE RE-SULTANT OR AMOUNT OF DIRECT FORCE OF AN UNIFORMLY VARYING FORCE DISTRIB-UTED OVER A CIRCULAR SEGMENT AREA.

Let Fig. 9 represent the pressure wedge, MBN the base or circular seg-ment, and B the origin of co-ordinates.

Let $x_c = AB$ the versine of one half of the arc MBN,

$x =$ any distance from the origin B,

$2 (2rx - x^2)^{\frac{1}{2}} dx =$ the differential of the segment,

$V =$ the resultant or volume of the pressure wedge,

$C =$ the greatest height, DB,

$$V = \int_0^{x_c} \frac{x_c - x}{x_c} C \times 2 (2rx - x^2)^{\frac{1}{2}} dx,$$

$$\therefore V = 2C \left[(2rx - x^2)^{\frac{1}{2}} \left(\frac{x_c + 3r^2 - 2r}{6} \right) + r \, versin.^{-1} \frac{x_c}{r} \left(\frac{r}{2} - \frac{r^2}{x_c} \right) \right].$$

Substituting the equalities given in Art. 12 and reducing, we have

$$V = \frac{C}{6d_c} \left[2 \sqrt{d_c d_\tau} \left[d^2_c + r (3r - 2d_c) \right] \right.$$

$$\left. + \textit{Segment arc } (d_c - r) 3r \right]. \ (17)$$

The resultant of a tension force may be obtained from the above equation by placing T for C.

The cubic contents of any cylindrical wedge may be com-puted from Eq. 17 by substituting for C, expressed in pounds, h, the greatest height of the wedge in inches.

The area of the base of the pressure wedge or that of any segment of a circle may be computed from the following :

$$Area = 2 \sqrt{d_c d_r}\,(d_c - r) + Segment\ arc \times \frac{d_c}{4}, \qquad (17A)$$

in which

d_c = versine of one half the arc of the segment,

r = the radius,

d_r = the diameter, less d_c.

SECTION V. — *Uniformly Varying Forces—Circular Arcs.*

15. Problem. REQUIRED THE MOMENT OF AN UNIFORMLY VARYING COMPRESSIVE FORCE DISTRIBUTED OVER THE ARC OF A CIRCLE, WITH RESPECT TO AN AXIS THAT IS A TANGENT TO THE CIRCLE AND PARALLEL TO THE CHORD JOINING THE EXTREMITIES OF THE ARC.

Let, in Fig. 10, OBS represent the axis, MBN the arc,' ABD a section through AB, B the origin of co-ordinates, C the intensity of the force at B, and zero that at M and N.

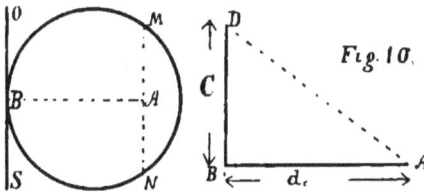

Fig. 10.

For *notation* refer to Arts. 7 and 12.

Let $x_c = d_c =$ the versine AB of one half the arc MBN,

$x =$ any variable distance from B or the axis,

$\dfrac{x_c - x}{x_c}$ C = the intensity of the force at x distance,

r = the radius of the circle,

$\dfrac{2r\,dx}{\sqrt{2rx - x^2}}$ = twice the differential of the arc,

$$R_c = \int_0^{x_c} \frac{2r\,dx}{\sqrt{2rx - x^2}}, \frac{(x_c - x)}{x_c} C x.$$

Integrating, we obtain

$$R_c = 2C\left[(2rx - x^2)^{\frac{1}{2}} \left(\frac{(x_c + 3r)}{2x_c} \, r - r \right) \right.$$
$$\left. + r\,versin.^{-1}\,\frac{x_c}{r}(r - \frac{3r^2}{2x_c}) \right].$$

Substituting the values given in Art. 12, we have

$$R_c = \frac{rC}{2d_c}\left[\sqrt{d_c d_\tau}\,[2(3r - d_c)] + Comp.\,arc\,(2d_c - 3r) \right]. \quad (18)$$

16. Problem. REQUIRED THE MOMENT OF AN UNIFORMLY VARYING FORCE DISTRIBUTED OVER THE ARC OF A CIRCLE, WITH RESPECT TO AN AXIS THAT IS A TANGENT TO THE CIRCLE, PARALLEL TO THE CHORD CONNECTING THE EXTREMITIES OF THE ARC, AND AT THE DISTANCE d_c FROM IT.

Let, in Fig. 11, FOS represent the axis, MBN the arc,

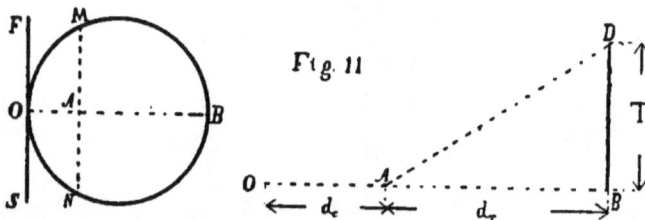

Fig. 11

$OABD$ a section through OB, T the intensity of the force at B, and zero at M and N.

For *notation* refer to Arts. 7 and 12.

Let $x_\tau = d_\tau =$ the versine of one half the arc MBN,

$x =$ any distance from the origin B,

$\dfrac{x_\tau - x}{x_\tau} T =$ the intensity of the force at any distance x,

$d - x =$ its lever-arm,

$\dfrac{2rdx}{\sqrt{2rx - x^2}} =$ the differential of the arc,

$$R_\tau = \int_0^{x_\tau} \frac{2rdx}{\sqrt{2rx - x^2}} \left(\frac{x_\tau - x}{x_\tau} \right) (d - x).$$

Integrating, we obtain

$$R_\tau = \frac{rT}{2d_\tau} \left[\sqrt{d_c d_\tau} \left[2 \left(r + d_\tau \right) \right] + Tension\ arc \left(2d_\tau - r \right) \right], (19)$$

from which the required moment may be computed.

17. Problem. REQUIRED THE RESULTANT OR AMOUNT OF
DIRECT FORCE OF AN UNIFORMLY VARYING FORCE DISTRIBUTED
OVER AN ARC OF A CIRCLE.

Let, in Fig. 9, page 16, MBN represent the arc of the circle, $MNBD$ a wedge whose cylindrical surface is equal to the required resultant, the force being C in intensity at B and zero at M and N.

Let $x_c = d_c =$ the distance AB,

$x =$ any distance from the origin of co-ordinates B,

$\dfrac{x_c - x}{x_c} C =$ the intensity of the force at the point x,

$\dfrac{2rdx}{\sqrt{2rx - x^2}} =$ the differential of the arc,

$V =$ the resultant or volume of the pressure wedge,

$$V = \int_0^{x_c} \frac{x_c - x}{x_c} \cdot Cx \cdot \frac{2rdx}{\sqrt{2rx - x^2}},$$

$$\therefore \; V = \frac{C}{d_c} \Big[2r \sqrt{d_c d_x} + \textit{Segment arc } (d_c - r) \Big], \qquad (20)$$

from which the required resultant may be computed.

The *curved surface* of a *cylindrical wedge* may be computed from the above formula by substituting for C, expressed in pounds, h, the greatest height of the wedge in inches.

CHAPTER II.

18. Moment of **Resistance.** The cross-section is the shape of the figure and the area that any material, such as a beam, would show, should it be ,cut into two pieces by a plane perpendicular to its length, and its *resistance* to *rupture* at this plane or section is the number of *inch-pounds* that its fibres will offer to forces tending to *cross-break* the beam or material of which it is a section : this is called the *Moment* of *Resistance* of the *cross-section*.

The *Moment* of *Resistance* varies in amount with the material and the shape of the cross-section, but it is entirely independent of the length of the beam and of the manner in which the load may be applied, in each case ; the same cross-section and material will offer the same number of *inch-pounds* of *resistance* when broken across. •

19. Neutral Line. When a beam is broken across, or is acted upon by forces that bend or tend to break it, we know

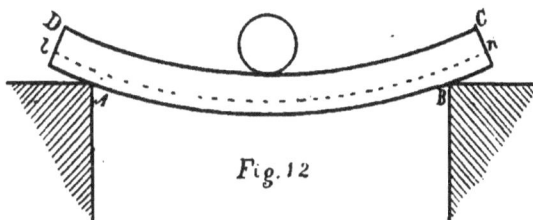

Fig. 12

from observation that its fibres on the lower or convex side, *AB* (Fig. 12), are in a state of tension, and that those on the upper or concave side, CD, are compressed ; but our knowledge obtained from observation is limited to what takes place on the surface of the beam. We can only know what takes place within such a beam by reasoning from analogy ; there is a tensile strain in the lower side of the beam, AB, and just the *reverse* character of strain in the upper side, CD. In order that these two *directly* opposite strains may exist in the same beam at the same time, both strains must decrease from the surface toward a common point within the beam, where both strains become zero in intensity, and they may be classed and treated as *uniformly varying* forces.

A line, nl, for the longitudinal section of the beam, or a plane for the beam, is called the *neutral line* or line of *no strain ;* its position in a beam having a cross-section of a given shape, at the instant of rupture, depends upon the material alone, or upon the *ratio* existing between the *breaking compressive* and *tensile fibre strains.* No line in this plane has any of the properties of an *axis* that are usually assigned to it by writers on this subject.

20. Bending Moment and **Moment** of **Resistance.** In order to obtain the relation existing between the *Bending Moment* of the applied load and the *Moment* of *Resistance* of the cross-section of the beam, conceive one half of the beam, $ABCD$ (Fig. 12), to be removed and the bent-lever, ogf (Fig. 13), to be substituted for it, and that the same *cohesion* to exist between the fibres of the bent-lever and those of the beam along the line, fy, that originally existed between the fibres of the two halves of the beam along the same line, the *bent-lever,* however, preserving its *distinctive* character of a *bent-lever.* The *applied load,* L, causes the bent-lever, ogf, and the half of the beam, $AgfD$, to move *downward* in the direction of the lower arrow of the figure, and the end of the lever, O, and that of the half of the beam, D, to *revolve around* f in the direction of the upper arrows,

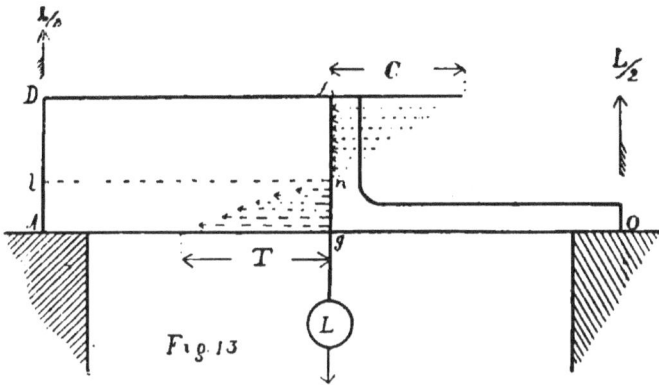

Fig. 13

the effect being identical with that produced by conceiving the *fulcrum*, f, to remain stationary and a moving force, $L \div 2$, to be applied to the end of the bent-lever, O, and at the same time to be pressed in the direction of its length by a force, C, equal in magnitude to the amount of direct compression along the line, fn, above the *neutral line*, nl. When the bent-lever is made to revolve around the fulcrum, f, it meets with an opposition of *compression* to its motion, *decreasing* in intensity from the *fulcrum*, f, to the *neutral line*, n, when it becomes zero in value; at this point the opposition changes to a *tensile* resistance which *increases* in intensity from zero at n to its maximum, T, at g.

The *bent-lever* and the *original vertical section* of the beam, fg, are pressed *closest* together at the *fulcrum*, f; from this point they continue to separate, by virtue of the ductility and compressibility of the material composing the beam, until *rupture* takes place, either in its upper or lower fibres; the action of the bent-lever being identical with that of the half of the beam for which it was substituted.

21. Equilibrium. The Bending Moment of the applied load, L, and the Moment of Resistance of the tensile and compressive fibre strains must be taken or computed with reference to the *fulcrum*, *f*, and in order that equilibrium shall exist, the Bending Moment of the applied load, *L*, must be equal to the Moment of Resistance, or the sum of the *moments* of the *tensile* and *compressive* fibre strains, and that the latter must be equal to each other in magnitude.

22. Position of the **Neutral Line.** By deducing general formulas for the *moments* of the *tensile* and *compressive* fibre strain in a cross-section of a given shape, placing them equal to each other and deducing from the equation so formed a general formula for either *gn* or *fn*, the *position* of the *neutral line* may be found in any section of the same shape by substituting in the formula its dimensions and any *known* values for the tensile and compressive fibre strains.

In sections of an *uniform shape*, such as the rectangle and the circle, the *depth* of the *neutral line* below the compressed side of the beam may be obtained by multiplying the depth of the rectangle and the diameter of the circle by a determined quantity that is constant for each shape and material; this constant multiplier being its position when the depth of the rectangle and diameter of the circle is *unity*.

But in *irregular shaped* sections, such as the T, Double T, Box, Rolled-eyebeam and other shapes, in which the metal is not continuous from the neutral line to the top and bottom, a special solution must be made for each case to determine the position of the *neutral line* from which to compute the Moment of Resistance of the section. Before beams of this character are manufactured, an economical position should be *assumed* and a sufficient area of metal placed above and below the *neutral line* to furnish the required Moment of Resistance.

23. Neutral Line at the **Transverse Elastic Limit.** Our object in *testing*, to destruction, the strength of any piece of construction material is to obtain information that will guide us to a correct knowledge of its use, when safety to life and property is demanded, whether this destruction be by means of *extension* or *compression ;* as we apply the load in small instalments there are only two points in its intensity at which we can record the knowledge thus gained for future intelligent, comparative use when it has reached the *elastic limit,* and when it is sufficiently intense to rupture or destroy the piece of material tested. When a beam is broken by a *transverse load* that has been applied in small instalments, we have two similar points, the *elastic limit* load and the *rupturing* load, and these are the only points at which our information can be used to compute the strength of similar material when used in structures.

We know that in an ordinary beam without a load its compressive and tensile fibre strain is zero, and that the *breaking* transverse load produces the *breaking compressive* and *tensile* strain in the fibres. Now, does the *transverse elastic limit load* produce the *elastic* fibre strain limits? Is the neutral surface the same as that for rupture? When the beam is unloaded each plane of fibres is a *neutral surface ;* as the load is applied the compressive and tensile strain penetrates the beam from the top and bottom respectively ; theoretically we know they must meet at a common point within the beam at the *instant* of *rupture,* and that this is fully sustained by experiments will be shown in the sequel. Our theory demands that when the same *ratio* exists between the tensile and compressive *elastic* fibre strain *limits* that does between the ultimate or breaking strains, in order that equilibrium shall exist, the *neutral surfaces* must be identical, but the theory does not require that the transverse elastic limit load shall produce the elastic fibre strain limits ; we can only gain the desired information from discussing a numerical example.

The *mean* compressive and tensile elastic fibre strain limits for good wrought-iron is $C = 30000$ pounds and $T = 30000$ pounds per square inch. With these strains, from formulas de-

duced in the sequel, the centre elastic limit transverse load of
a bar of wrought-iron six inches square and ten feet span is
44100 pounds. The centre elastic limit transverse load of a
bar of wrought-iron, one inch square and one foot span, is
2250 pounds, and that of the above beam from the well-
known formula is,

$$\text{Load} = \frac{bd^2S}{l} = \frac{6 \times 36 \times 2250}{10} = 48600 \text{ pounds.}$$

From this practical identity of results, as we have only used
average values, and other special tests given in the sequel, we
are authorized to conclude that the *transverse elastic limit
load* produces the *tensile* and *compressive* fibre strain *elastic
limits*.

24. Movement of the **Neutral Line** with the **Deflection.** Having established the fact that the transverse
elastic limit load produces the elastic fibre strain limits—and
our theory requires that the elastic limit neutral line and the
neutral line of rupture shall coincide only when their ratios
are the same, but should they be unequal they must occupy
different positions—in the sequel it will be shown that where
these ratios are unequal the elastic limit neutral line is situated
between the neutral line of rupture and the bottom or extended side of the beam, and that as the loading advanced
from the elastic limit load to the rupturing load, the neutral
surface must have *moved upward* or toward the compressed
surface of the beam.

From the above we conclude that as there was no change
in the condition of the loading that could have *reversed* the
direction in which the neutral line *moved* from its position at
the *elastic limit* to that at *rupture*, the neutral line at the inception of the loading was at the bottom or extended side of
the beam, and that as the loading progressed it *moved upward* or toward the compressed side of the beam—the tension
area, to avoid rupture in its fibres, continues to encroach upon

the compressed area until the rupturing strain is produced in both the top and the bottom of the beam.

The *neutral line*, at the inception of the loading, being at the bottom or extended side of the beam, it can only be moved upward by reason of the *deflection* and equally with it. If, from the dimensions of the beam, it should not be able to deflect sufficiently to move the neutral line to the position required for *equilibrium* between moments of resistance of the *ultimate* fibre strains, the true breaking strength will not be obtained for the beam. When this is the case the observed breaking load will be too large for wooden, wrought-iron, steel and tough cast-iron beams, and too small for the more fractious varieties of material; for should the compressive strain reach its ultimate limit before the tensile strain an increase of the load will develop a crushing strain in excess of the true crushing intensity, as is frequently done in crushing short blocks; the beam will, however, continue to deflect under these increased loads, and will finally develop the full tensile strength, when the beam will be broken by a load much in excess of its true breaking load.

From the above, the reason for the variation in the *modulus* of *rupture* that is required in the "common theory of flexure" is apparent, as the shorter beams in most series of experiments, especially of cast-iron, did not deflect sufficiently to break with the true breaking load, and, therefore, it required a larger *modulus* or empirical coefficient for the shorter beams than for the longer beams of the same series of tests.

At the *instant* of deflection the *bending moment* of the applied load is held in equilibrium by a *purely* compressive resistance, distributed over the section as an *uniformly varying* force, being zero in intensity at the bottom or extended side of the beam and greatest in intensity on the opposite side. This is a very important principle, as from it we shall, in the sequel, deduce the correct theory of the strength of columns.

25. Neutral Lines of Rupture in Rectangular Sections. It may not be out of place here to anticipate the results obtained in the sequel, by stating the positions of the neutral lines of rupture in a rectangular section when composed of the different kinds of material used in construction.

Let the section be six inches deep, q the ratio of the compressive to the tensile strains of rupture, or $C \div T$.

Cast-iron........$q = 8.5$	⎧ depth of neutral line below the compressed side of the section......... ⎫	2.45 ins.
" ......$q = 5.$	" "	3.00 "
" ......$q = 4.$	" "	3.24 "
Steel...........$q = 1.5$	" "	4.29 "
Wrought-iron ....$q = 1.$	" "	4.68 "
Beech, English....$q = 0.775$	" "	4.90 "
" American $q = 0.383$	" "	5.36 "

From this comparative statement we observe the order in which the neutral lines of *rupture* are arranged in rectangular sections; the same order of arrangement exists in all other uniform sections. For different kinds of wood and cast-iron the neutral line of rupture lies between the extremes given in the above table of comparison.

26. Relative Value of the Compressive and Tensile Strains. Experiments have fully shown that the *compressive* and *tensile* strains do not possess equal values as factors in determining the transverse load that a beam will bear, and that the influence of the tensile strain predominates.

The *elastic limit* or *technical breaking* load of a wrought-iron beam one inch square and twelve inches span, loaded in the middle, is 2000 pounds when

$C = 30000$, $T = 30000$ pounds, and the Moment of Resistance $= 6084$ inch-pounds from our formulas.

We will now endeavor to trace the effect that will be produced upon the amount of the transverse breaking load from varying the *exact* and *relative values* of C and T from those in the above-described wrought-iron beam, which will be taken as our standard of comparison. Since the crushing and tensile strength of the material composing any beam must each sustain one half of the breaking load of the beam, the *per cent* of loss or gain in the transverse strength should be, *approximately,* one half the sum of the per cents of the losses or gains in the values of C and T, but this will be found to be true for only the smaller ratios of $C \div T$ that exist in materials of construction.

The centre breaking load of a white pine beam one inch square and twelve inches span is 450 pounds when

$C = 5000$, $T = 10000$ pounds, and the Moment of Resistance $= 1260$ inch-pounds. In passing from the standard wrought-iron beam where $C = T$ to the white pine beam where $C = 0.5$ T, the following changes take place :

Loss in the value of C..................... 83.4 per cent,
"   " T..................... 66.6   "

———

Apparent loss to the transverse load.......... 75.0   "
Actual   "   "   "   .... 77.5   "
Loss to the Moment of Resistance........... 79.2   "

results, practically, identical for this ratio.

From Mr. Kirkaldy's experiments a bar of steel one inch square and twelve inches span will break with a centre load of 6400 pounds when

$C = 160000$, $T = 70000$ pounds, Moment of Resistance $= 19200$ inch-pounds. In passing from our standard wrought-iron beam to the steel beam, the following changes take place :

Gain in the value of C.................... ....433.3 per cent.
"   " T.... ....133.3   "

Apparent gain to the transverse load.........283.3   "
Actual   "   "   "   " ....220.0   "
Gain to the Moment of Resistance......... ..215.0   "

Mr. Hodgkinson found the centre breaking load of a certain
cast-iron beam, one inch square and twelve inches span, to be
2000 pounds when
$C = 115000$, $T = 14200$ pounds, and the Moment of Resistance $= 6600$ inch-pounds. In passing from our standard
wrought-iron to the cast-iron beam of the same size, the
following changes take place:

Gain in the value of C.................. ... $+$ 283.4 per cent.
Loss "   "   "   " T................. $-$ 52.6 "   "

Apparent gain to the transverse load ....   115.4 "   "
Actual   "   "   "   " .....   0.0 "   "
Gain to the Moment of Resistance ....   8.4 "   "

In this experiment it required 283.4 per cent gain in the
value of C to offset a loss of 52.6 per cent in the value of T,
or that the compressive strength does not sustain its proper
proportion of the load.

This great discrepancy between the legitimate theoretical
deductions and the results obtained from experiments cannot
be reconciled on the hypothesis that the forces are in equilibrium with respect to an *axis* that lies within the beam,
the moment in each case, for rectangular sections, being the
resultants of the tensile and compressive forces, multiplied by
two thirds of the depth of their respective areas, showing that
the compressive strain *works* under no disadvantages ; but on
our theory this discrepancy is fully accounted for.   The lever-
arm of the *crushing* resultant is one third the depth of the

compressed area, while that of the *tensile* resultant is only one third less than the total depth of rectangular beams.

27. Summary of the Theory. The theory herein advanced to explain the relation that exists between the Bending Moment of the applied load and the Moment of Resistance of the material composing the beam, may be expressed by the following hypotheses :

1st. The fibres of the beam on its convex side are extended and those on the concave side are compressed in the direction of the length of the beam, and there are no strains but those of extension and compression.

2d. There is a layer or plane of fibres between the extended and compressed sides of the beam that is neither extended nor compressed, which is called the *neutral surface* or *neutral line* for any line in this plane.

3d. The strains of compression and extension in the fibres of the beam are, in intensity, directly proportional to their distance from the neutral surface.

4th. The axis or origin of moments for the tensile and compressive resistance of the fibres of any section at right angles to length of the beam, is a line of the section at its intersection with the top or compressed side of the beam.

5th. The fibres of a beam will be ruptured by either the tensile or compressive strains in its concave and convex surfaces, whenever they reach in intensity those found by experiment to be the direct breaking tensile and compressive fibre strains for the material composing the beam.

6th. The Bending Moment of the load at any section is equal to the sum of the moments of resistance to compression and extension of the fibres, or to the Moment of Resistance of the section of the beam.

7th. The sum of the moments of resistance of the fibres to compression is equal to the sum of the moments of resistance of the fibres to extension.

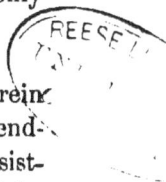

8th. The algebraic sum of the direct forces of compression and extension can never become zero.

9th. The Moment of Resistance of the section is equal to the sum of the moments of resistance to the compression and extension of its fibres.

10th. The transverse elastic limit load produces the tensile and compressive fibre strain elastic limits.

CHAPTER III.

SECTION I.—*General Conditions.*

28. Coefficients or **Moduli** of **Strength** are quantities expressing the intensity of the strain under which a piece
of a given material *gives way* when strained in a given
manner, such intensity being expressed in units of weight for
each unit of sectional area of the material over which the
strain is distributed. The *unit* of weight ordinarily employed
in expressing the strength of materials is the *number of pounds
avoirdupois* on the *square inch.*

Coefficients of *Strength* are of as many different kinds as
there are different ways of breaking a piece of material.

·*Coefficients* of *Tensile Strength* or *Tenacity* is the strain
necessary to *rupture* or pull apart a prismatic bar of any given
material whose section is one square inch, when acting in the
direction of the length of the bar. This strain is the *T* of
our formulas.

Coefficient of *Crushing Strength* or *Compression* is the
pressure required to *crush* a prism of a given material whose
section is one square inch, and whose length does not exceed
from *one* to *five times* its diameter, in order that there may be
no tendency to give way by bending sideways. This pressure
is the *C* of our formulas.

29. Elasticity of **Materials.** It is found by experiment that if the load necessary to produce a strain and fracture
of a given kind is applied in small instalments, that before
the load becomes sufficiently *intense* to produce rupture, it will

cause a change to take place in the form of the material, and if the load is removed before this intensity of the fibre strain passes *certain* limits, the material possesses the power of returning to its original form. This is called its *elasticity*.

30. Elastic Limits. When the material possesses the power of recovering its exact original form without " set" on the removal of a load of a given intensity, the greatest load under which it will do this is called the *limit* of *perfect elasticity*.

The *limit* of *elasticity* as ordinarily defined and used by experimenters is that point or intensity of strain where equal instalments or increments of the applied load cease to produce equal changes of form, or where the change in form increases more rapidly than the load.

31. The Elastic Limit of Beams may be determined by applying small equal parts of the load and noting the increase in deflection after each increase of the load, allowing sufficient time for each increase of the load to produce its full effect. When it is found that the deflections increase more rapidly than the load, its *elastic limit* has been reached and passed. The relation between the elastic limit load of the beam and the elastic limit of the tensile and compressive fibre strains of the material composing the beam will be shown in the sequel, or that the *elastic limit load* of the beam produces the *elastic limit* strain for the fibres.

32. Working Load and Factor of Safety. The greatest load that any piece of material, used in a structure, is expected to bear is called the *working load*.

The *breaking* load to be provided for in *designing* a piece of material to be used in a *structure* is made greater than the *working* load in a certain ratio that is determined from experience, in order to provide for unforeseen defects in the material and a possible increase in the magnitude of the expected working load.

The *factor* of *safety* is the ratio or quotient obtained by dividing the *breaking* load by the *working* load required.

33. General Formula. In our first chapter we deduced rules or formulas, from which can be computed the Greatest Bending Moment that a load applied to a beam in a given manner will produce without reference to the shape of its cross-section, or to the material composing the beam.

In our second chapter *principles* are deduced from which can be computed the Greatest Moment of Resistance cross-sections of the various shapes and material will exert at the instant of rupture, without reference to the length of the beam or to the manner in which the load may be applied.

To avoid repetition, the formulas for the Moments of Resistance are deduced in this chapter. Our knowledge of the *transverse strength* of beams will now be complete if we compute and make the Greatest Moment of Resistance of the cross-section of the beam equal to the Greatest Bending Moment of the applied load.

Let $R =$ the Greatest Moment of Resistance of the beam,

$L =$ the total applied load in pounds,

$s =$ the span, the distance between the supports in inches,

$n =$ the factor defined in Art. 6, page 7,

$M =$ the Greatest Bending Moment of the applied load.

From Eq. 8, page 7, we have

$$M = nLs = R,$$

$$\therefore \; L = \frac{R}{ns}, \qquad (21)$$

from which the breaking load of the beam may be computed.

34. Relative Transverse Strength of a Beam. Referring to the values of the factor, n, of Eq. 21, as given in Art. 6, page 7, it is found in each case to be either *unity* or a *fraction*, and when these values are introduced for n, in Eq. 21, it is equivalent to multiplying the numerator, R, by the denominator of the fraction, n; hence if we make

$$\frac{1}{n} = m,$$

Eq. 21 will become

$$L = \frac{m R}{s}. \qquad (22)$$

Computing and tabulating the values of m from those of n, Art. 6, we obtain the following relation between the breaking loads of a beam, the *span*, *material* and *cross-section* remaining the same or *constant;* the breaking load of a beam, fixed at one end and loaded at the other, being the unit or standard of strength.

$$m$$

Beam fixed at one end and loaded at the other............ 1
Beam   "   "   "   "   uniformly.... 2
Beam supported at its ends and loaded at its middle....... 4
Beam   "   "   "   "   "   uniformly......... 8
Beam fixed at both ends and loaded at its middle......... 8
Beam   "   "   "   "   "   uniformly...........12

Eq. 22 is the general formula from which will be deduced the transverse strength of beams of all sections by giving to the factor, R, its proper value.

SECTION II.—*Transverse Strength—Rectangular Sections.*

35. Moment of Resistance.

In Fig. 14, let $ABDX$ represent the section, nl the neutral line, $AnlX$ the *compressed* area, $BnlD$ the *extended* area, $AnBMnNA$ any section through

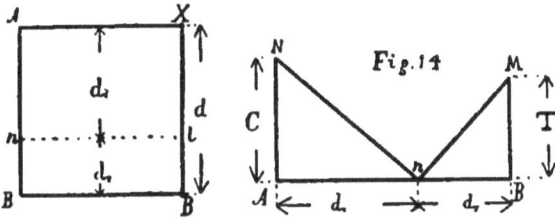

Fig. 14

the strain wedges at right angles to the neutral line, and AX the *axis* or origin. Adopting the *notation* heretofore used in Art. 7, we have

C = the greatest intensity, per square inch, in pounds of the compressive strain,

T = the greatest intensity, per square inch, in· pounds of the tensile strain,

R_c = the moment of the compressive strain in inch-pounds,

R_τ = the moment of the tensile strain in inch-pounds,

$R = R_c + R_\tau$ = the Moment of Resistance of the section in inch pounds,

d_c = the depth of the compressed area, An, in inches,

d_τ = the depth of the extended area, Bn, in inches,

$d = d_c + d_\tau$ = the depth of the section AB in inches,

b = the width of the section BD in inches,

q = the quotient arising from dividing C by T,

L = the total applied load in pounds,

s = the span, the distance between the supports in inches,

m = the factor, defined in Art. 34, page 35.

From Eq. 11, page 10, we have, for the moment of compressive resistance,

$$R_c = \frac{bd_c^2 C}{6},$$ (23)

and for the moment of the tensile resistance from Eq. 13,

$$R_\tau = \frac{bd_\tau}{6}(2d + d_c)T.$$ (24)

36. The Neutral Line. From the 7th hypothesis, Art. 23, page 31, we have, from equating the second members of Eqs. 23 and 24,

$$\frac{bd_c^2 C}{6} = \frac{bd_\tau}{6}(2d + d_c)T,$$

from which

$$d_\tau = \frac{d(q + 1\cdot5 - \sqrt{2q + 2\cdot25})}{q + 1},$$ (25)

and

$$d_c = \frac{d(-0\cdot5 + \sqrt{2q + 2\cdot25})}{q + 1}.$$ (26)

The position of the *neutral line* in any rectangular section may be computed from either of these formulas; it is independent of the absolute intensity of the maximum tensile and compressive fibre strains, but depends upon their ratio $C \div T = q$ for its position. The application of these formulas is illustrated in Examples 5, 6 and 7 of the sequel.

37. Transverse Strength. General formulas for the *transverse strength* may be obtained by substituting for R the Moment of Resistance in Eq. 22, page 36, its values $2R_c = 2R_\tau$ from Eqs. 23 and 24.

$$\therefore L = \frac{mbd_c^2 C}{3s},$$ (27)

$$\therefore L = \frac{mbd_\tau}{3s}(2d + d_c)T.$$ (28)

From either of these equations, either the breaking load or the elastic limit load may be computed by giving to C and T the *breaking* or *elastic limit* values of the material composing the beam. The application of Eq. 27 is illustrated in Examples 7, 26, 27, 28, 36, 37, 38, 39 and 40, and Eq. 28 in Examples 5, 6, 7, 8, 25 and 30 of the sequel.

38. To Design a Beam. In designing a beam of a rectangular section that shall break with a given total applied load, distributed over the span of the beam in a given manner, the span of the beam will be determined from the position in which it is to be used, and the depth, d, in inches will be *assumed*; the *crushing* and *tensile* strength of the material composing the beam must be *known*; it will then only be necessary to determine the *width*, b, in order that the beam shall break with the required load.

From Eq. 25 deduce the position of the *neutral line*, which is independent of the width, b, then from Eq. 27 deduce the value of b, the width

$$\therefore b = \frac{3Ls}{md_c^2 C}, \tag{29}$$

which gives the required width, since all of the factors in the second member of this equation are known quantities.

Should the *assumed* depth and the *computed* width not give an economical section for the beam, a second depth must be assumed from this information and a second width computed; this process should be repeated until a satisfactory result is secured.

39. To Compute the Compressive and Tensile Strains.

PROBLEM I.—*The position of the neutral line and the crushing strain of the material of a rectangular section may be computed from the known transverse breaking load and tensile strength.*

From Eq. 28, page 38, deduce the value of d_τ,

$$d_\tau = \frac{3d}{2} - \sqrt{\frac{9mbd^2T - 12Ls}{4mbT}}, \tag{30}$$

and

$$d_c = d - d_\tau,$$

which gives the required position of the neutral line, illustrated in Examples 1, 2, 21, 22 and 35.

The *crushing* strength can be computed by deducing its value, C, from Eq. 27,

$$\therefore C = \frac{3Ls}{mbd_c^2}. \tag{31}$$

The application is illustrated in Examples 1, 2 and 21.

PROBLEM II.—*The position of the neutral line and the tensile strength of the material of a rectangular section may be computed from the known transverse breaking load of the beam and the crushing strength of the material.*

From Eq. 27 we have

$$d_c = \sqrt{\frac{3Ls}{mbC}}. \tag{32}$$

for the position of the *neutral line*, which is illustrated in Example 4.

From Eq. 28 we have for the required tensile strength

$$T = \frac{3Ls}{mbd_\tau(2d + d_c)}, \tag{33}$$

illustrated in Example 4.

SECTION III.—*Transverse Strength—Hodgkinson Section.*

40. Moment of Resistance. In Fig. 15 let $ABDX$ represent the section, AOX the *axis* or origin of moments,

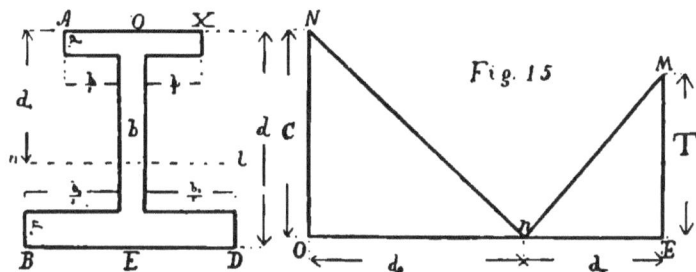

Fig. 15

nl the *neutral line*, the area above it being *compressed* and that below it *extended*.

On $EMnNO$, a section through the strain wedges on the line OE.

d = the depth of the web and section, OE, in inches,
d_1 = the " " upper flanges in inches,
d_2 = the " " lower " "
b = the width of the web in inches,
b_1 = the sum of the widths of the upper flanges in inches, or $AX - b$,
b_2 = the sum of the widths of the lower flanges in inches, or $BD - b$.

For other notation refer to Art. 35.

From Eqs. 11 and 12, page 10, we have for the *moments* of *compressive resistance*,

$$\text{Web, } R_c = \frac{bd_c^2 C}{6},$$

$$\text{Upper flanges, } R_c = \frac{b_1 d_1^2}{6d_c}(3d - 2d_1)\, C.$$

Adding, we obtain for the section,

$$R_c = \left[\frac{bd_c^2 + b_1 d_1^2 (3d_c - 2d_1)}{6d_c} \right] C, \qquad (34)$$

and for the *moments* of *tensile resistance* from Eqs. 13 and 14, page 11,

$$\text{Web, } R_\tau = \frac{bd_\tau}{6} (2d + d_c) \, T,$$

$$\text{Lower flanges, } R_\tau = \frac{b_2}{6d_\tau} \left[3d_2 d_\tau (2d - d_2) - d_2^2 (3d - 2d_2) \right] T.$$

Adding the above equations we obtain for the *moment* of *tensile* resistance of the section,

$$R_\tau = \left[bd_\tau^2 (2d + d_c) + b_2 [3d_2 d_\tau (2d - d_2) - d_2^2 (3d - 2d_2)] \right] \frac{T}{6d_\tau} \quad (35)$$

and the Moment of Resistance of the Hodgkinson section is $2R_c = 2R_\tau$ from Eqs. 34 and 35.

41. The Neutral Line. By equating the values of R_c and R_τ given by Eqs. 34 and 35, and deducing from the equation so formed the value of d_c, the position of the *neutral line* is determined, but as this will involve the solution of a biquadratic equation, the general solution will be too complex for ordinary use.

The following approximate formula obtained by neglecting certain quantities that do not materially affect the result in ordinary cases, will give its position sufficiently near for all practical purposes when the beam is made of cast-iron:

$$d_c = \frac{db + \sqrt{12bdd_2 b_2 (q + 1) + b^2 d^2 (4q + 5)}}{2b (q + 1)}. \qquad (36)$$

Its application is illustrated in Examples 8, 9, 10 and 29.

The position of the *neutral line* in a Hodgkinson or in any single or double flanged beam may be computed when the transverse strength of the flanged beam and either the compressive or tensile strength of the material have been determined experimentally.

PROBLEM I.—*Required the position of the neutral line in any flanged beam when the transverse, ultimate, or elastic load and the corresponding compressive strength of the material have been obtained from experiments.*

For any given flanged beam with its transverse and compressive strength determined experimentally, all of the terms of Eq. 37, page 44, except d_c, become known quantities; by giving the letters their numerical values for any beam, the formula can always be reduced to the following general form:

$$d_c^3 \pm 3pd_c \pm 2k = 0,$$

in which $3p$ and $2k$ are numerical quantities; then by Cardan's Rule for the solution of cubic equations of this form,

$$d_c = \sqrt[3]{k + \sqrt{k^2 + p^3}} + \sqrt[3]{k - \sqrt{k^2 + p^3}}, \quad \text{(A 36)}$$

p and k must be made equal to one third and one half of the numerical quantities, $3p$ and $2k$, respectively, and their algebraic sign must be that of the term in which they appear in the reduced equation.

From the above equation one value of d_c will be determined, and with it, Eq. (A 36) may be reduced to an equation of the *second* degree by dividing the cubic equation by d_c plus or minus the numerical value deduced above, giving the numerical value in the division the reverse sign to that computed. From this resulting equation of the second degree, the other two values of d_c may be computed; the value that represents the position of the neutral line may be generally determined from inspection.

PROBLEM II.—*Required the position of the neutral line in any flanged beam when the transverse load and tensile strength of the material have been obtained from experiments.*

In Eq. 38, which gives the relation between the Moment of

the applied load and the Moment of Resistance of the section, all of the terms become known quantities except d_τ; substituting for the letters their values in any given case, it will always reduce to the following general form:

$$d_\tau^{\,3} - a d_\tau^{\,2} \pm m d_\tau \pm n = 0,$$

in which a, m, and n are numerical quantities. By substituting for d_τ a new unknown quantity,

$$d_\tau = x + \frac{a}{3},$$

the equation will reduce to the following general form:

$$x^3 \pm 3px \pm 2k = 0, \qquad \text{(B 36)}$$

which can be solved by Cardan's Rule, as in the preceding Problem.

42. Transverse Strength. To obtain a general formula from which the *transverse strength* of a Hodgkinson beam may be computed, place for R, the Moment of Resistance in Eq. 22, page 36, its values, $2R_c = 2R_\tau$, from Eqs. 34 and 35, and we have

$$L = \frac{mC}{3 d_c s}\Big[b d_c^{\,3} + b_1 d_1^{\,2}\,(3 d_c - 2 d_1) \Big], \qquad (37)$$

and

$$L = \frac{mT}{3 d_\tau s}\Big[b d_\tau^{\,2}\,(2d + d_c) + b_2\,[3 d_2 d_\tau\,(2d - d_2) - d_2^{\,2}\,(3d - 2d_2)] \Big] (38)$$

From either of these formulas, either the breaking or elastic limit load may be computed. Their application is illustrated in Examples 8, 9, 10, 22, 29, 30 and 31.

43. To Design a Hodgkinson Beam. An economical position for the neutral line must be assumed and a sufficient area of the metal arranged above and below it, to furnish the required *compressive* and *tensile resistance ;* the depth of the beam, d, and the thickness of the web, b, should also be assumed.

Let $L =$ the required breaking load in pounds,
$s =$ the span in inches,
$m =$ the factor defined in Art. 35.

Step I.—Assume the thickness of the web, b, and its depth, d, which is also the depth of the beam, and an economical position for the neutral line. The *compressed* area above the neutral line must furnish *one half* and the *extended* area below it the *other half* of the required strength of the beam.

FOR THE COMPRESSED AREA:

Step II.—Place the value of the compressive resistance given by Eq. 11, page 10, for R in Eq. 22, page 36, and we shall have for the load, $L_{\text{\tiny 1}}$, sustained by the *compressed area* of the *web*,

$$L_{\text{\tiny 1}} = \frac{mbd_c^2 C}{6s}. \tag{39}$$

Step III.—For the *top flanges* deduct the load, $L_{\text{\tiny 1}}$, obtained in Step II., from one half of the required breaking load, L, of the beam ; the balance is the breaking load, $L_{\text{\tiny 2}}$, for the top flanges, or,

$$L_{\text{\tiny 2}} = \frac{L}{2} - L_{\text{\tiny 1}}. \tag{40}$$

In order to *design* an area of section that will sustain this load, we must assume a convenient depth, $d_{\text{\tiny 1}}$, for the top flanges and obtain the *sum* of their widths, $b_{\text{\tiny 1}}$, by placing for R, in Eq. 22, page 36, its value for this case, from Eq. 12, page 10, and we will have

$$L_{\text{\tiny 2}} = \frac{b_{\text{\tiny 1}} d_{\text{\tiny 1}}^2 (3d_c - 2d_{\text{\tiny 1}})\, C}{6d_c s}, \tag{41}$$

from which we have

$$b_{\text{\tiny 1}} = \frac{6d_c Ls}{d_{\text{\tiny 1}}^2 (3d_c - 2d_{\text{\tiny 1}})\, mC} \tag{42}$$

One half of this computed width, b_1, must be arranged on each side of the web.

FOR THE EXTENDED AREA:

Step IV.—Substitute for R, in Eq. 22, page 36, its value from Eq. 13, page 11, and we will have for the load, L_3, sustained by the *extended* area of the web,

$$L_3 = b d_\tau (2d + d_c) \frac{mT}{6s}. \qquad (43)$$

Step V.—For the *bottom flanges* deduct the load, L_3, found by Step IV., from *one half* of the required breaking load, L, of the beam; the remainder will be the load, L_4, that must be sustained by the bottom flanges, or

$$L_4 = \frac{L}{2} - L_3. \qquad (44)$$

In order to *design* an area for the *bottom flanges* that will sustain this load L_4, we must assume a convenient depth, d_2, for the bottom flanges, and obtain the sum of their widths, b_2, by substituting for R, in Eq. 22, its value for this case from Eq. 14, page 12, and we will have

$$L_4 = b_2 \left[3d_2 d_\tau (2d - d_2) - d_2{}^2(3d - 2d_2) \right] \frac{mT}{6d_\tau s}, \qquad (45)$$

from which we deduce

$$b_2 = \frac{6 d_\tau L_4 s}{[3d_2 d_\tau (2d - d_2) - d_2{}^2(3d - 2d_2)] mT}. \qquad (46)$$

One half the width computed from this formula must be arranged on *each* side of the web.

Step VI.—Should the *computed* dimensions from those *assumed* produce a badly designed section, new dimensions must be assumed from the knowledge thus obtained, and a second computation made as in the first instance.

SECTION IV.—*Transverse Strength—Double T and Hollow Rectangle or Box Section.*

44. Moment of **Resistance.** Let Fig. 16 represent the sections of the Double T and Box Sections respectively ; the . same letters in the text apply to both sections.

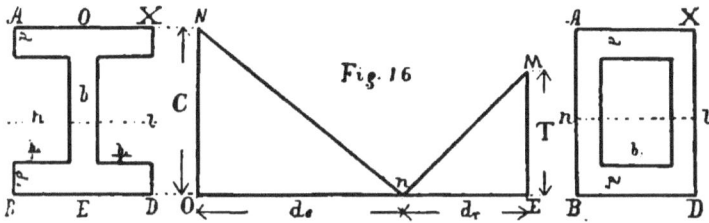

Fig. 16

Let $ABDX$ represent the sections, $OnEMnNO$ a section through the strain wedge on the line OE, AOX the axes, and nl the neutral lines.

Notation for the Double T :

  $b =$ the width of the web in inches,
  $b_1 = b_2 =$ the sum of the widths of the flanges of the double T in inches, or $AX - b$,
  $d_1 =$ the depth of the top flanges of the double T,
  $d_2 =$ the   "   "   " bottom "   "   "   "   "

Notation for the Box Section :

  $b =$ the sum of the widths of the sides of the box in inches,
  $b_1 = b_2 =$ the inside width of the box,
  $d_1 =$ the depth of the top of the box,
  $d_2 =$ the   "   "   " bottom of the box.

Giving the letters the above definition the *moments* of *tensile* and *compressive resistance* may be obtained from the following formulas :

$$R_c \text{ from Eq. 34,} \tag{47}$$

$$R_\tau \text{ `` `` 35.} \tag{48}$$

45. Neutral Line. The position of the neutral line in the Double T and Box Sections, when constructed of cast-iron, may be computed *approximately* from the formula given for the neutral line in the Hodgkinson beam, or

$$d_c \text{ from Eq. 36.} \tag{49}$$

When beams of these sections are made of wrought-iron and steel, the neutral line is so close to the bottom of the section that its position cannot be determined from an approximate formula. An exact solution of a long equation of the fourth degree must be made to determine its position.

46. Transverse Strength. The breaking strength, *L*, may be computed from the formula given for the transverse strength of the Hodgkinson beam, by noting the definition of the letters given in Arts. 35 and 44.

$$L \text{ from Eq. 37,} \tag{50}$$

$$L \text{ `` `` 38.} \tag{51}$$

47. To Design a Double T and Box Section.

Step I. *Assume* the *depth, d,* of the Box or Double T and the position of the *neutral line,* also the width, *b,* of the web of the Double T, or the sum of the equal widths of the sides of the box. The *compressed* area above the assumed position of the neutral line must sustain *one half* and the *extended* area below it the other *half* of the Bending Moment of the applied load, *L,* that the beam is required to break with.

FOR THE COMPRESSED AREA :

Step II.—The moment of compressive resistance that the sides of the Box or the web of the Double T will offer to the

Bending Moment of the applied load from Eq. 11, page 10, being substituted for R in Eq. 22, page 36, will give for its proportion of the load

$$L_{\iota} = \frac{mbd_c^2C}{6s}. \tag{52}$$

Step III.—For the *top flanges* of the Double T or the *top* of the Box Section deduct the load, L_ι, found by Step II. from *one half* of the total applied load, L; the remainder will be the breaking load, L_2, for the top flanges or the top of the box, as the case may require, or

$$L_2 = \frac{L}{2} - L_\iota. \tag{53}$$

To *design* an area of section that will resist this load, L_2, assume a width, b_ι, for the sum of the widths of the top flanges or the top of the Box, and compute therefrom their depth, d_ι, by substituting for the *moment* of *compressive resistance*, R, in Eq. 22, page 36, its value in this case from Eq. 12, page 10, and we will have

$$L_2 = b_\iota d_\iota^2 (3d_c - 2d_\iota) \frac{mC}{6d_c s}, \tag{54}$$

from which deducing the value of d_ι by making $d_\iota^2 = d_\iota^3$, which may be done without appreciably affecting the result obtained, we have

$$d_\iota = \sqrt{\frac{6d_c L_2 s}{b_\iota (3d_c - 2) mC}}, \tag{55}$$

from which the required depth may be computed.

FOR THE EXTENDED AREA:

Step IV.—The *moment* of the *tensile* resistance that the web of the Double T or the sides of the Box Section will offer to the breaking moment of the applied load, L, will be obtained from Eq. 13, page 11, which being placed for R in Eq. 22, page 36, gives for its proportion of the load, L_3,

$$L_3 = bd_\tau (2d + d_c) \frac{mT}{6s}. \qquad (56)$$

Step V.—The *bottom flanges* of the Double T or the *bottom* of the Box Section must sustain as its proportion of the load, L_4,

$$L_4 = \frac{L}{2} - L_3. \qquad \cdot \qquad (57)$$

In order to *design* an area that will sustain this load, L_4, assume a width, $b_2 = b_1$, the width assumed in Step III., and compute the corresponding depth, d_2, by placing for R, in Eq. 22, the value of the moment of the tensile resistance for this case, from Eq. 14, page 12, and we have

$$L_4 = \left[3d_2 d_\tau (2d - d_2) - d_2^2 (3d - 2d_2) \right] \frac{mb_2 T}{6d_\tau s}, \qquad (58)$$

from which, deducing the value of d_2, by making $d_2^2 = d_2^3$, we have

$$d_2 = \frac{3b_2 d_\tau m T + \sqrt{(3b_2 d_\tau m_\tau T)^2 - 24b_2^2 d_\tau L_4 s \, (mT)^2 (3d_\tau + 3d + 2)^2}}{2b_2 m T (3d_\tau + 3d + 2)}, \qquad (59)$$

from which the required depth may be computed.

Step VI.—Should the section obtained by this process not be satisfactory new dimensions must be assumed to remedy the defects of form, and the process repeated.

CAST-IRON DOUBLE T AND BOX SECTIONS:

For the Box or Hollow rectangular beam the assumed widths, $b_1 = b_2$, must be placed between the sides of the box, that having the depth, d_1, from Eq. 55, must be placed at the top, and d_2, computed from Eq. 59, at the bottom of the Box beam.

For the Double T beam the assumed width, $b_1 = b_2$, must be placed in equal projecting flanges on each side of the web at the top and bottom, that having the depth d_1 at the top and d_2 at the bottom of the beam.

WROUGHT-IRON DOUBLE T AND BOX SECTION.

For the Box or Hollow rectangular beam made of riveted plates a *part* of the assumed width, $b_1 = b_2$, must be placed between the sides of the Box and the balance in two equal projecting flanges on the outside of the Box at the top and bottom; the *total* width of the rolled plate that forms the top and bottom of the Box is equal to $(b + b_1)$, and the depth of the plates that form its sides is equal to $d - (d_1 + d_2)$ as found by Steps III. and V.

For the Double T:

Riveted plate sections. The width of the top and bottom plates is equal to $(b + b_1)$, and the depth of the web plate is $d - (d_1 + d_2)$.

Rolled Eyebeams are arranged like that for a cast-iron Double T beam.

SECTION V.—*Transverse Strength.*

The Inverted T, Double Inverted T and ⊔ Sections.

48. Moment of **Resistance.** Let Figs. 17, 18 and 19 respectively represent the three sections, *nl* the *neutral lines*, and *AX* the *axes* or origin of moments of resistance for each section.

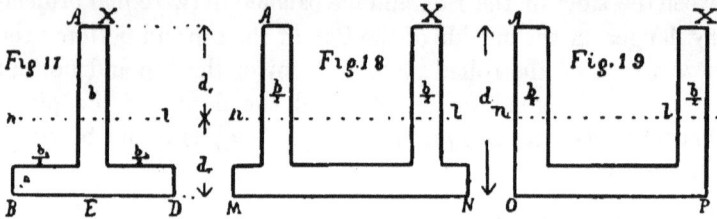

Fig 17   Fig. 18   Fig. 19

In addition to the *notation* given in Art. 35, observe the following for use in the formulas given in this Section:

The Inverted T, Fig. 17.

b = the width of the web in inches,

b_2 = the sum of the widths of the flanges, or $BD - b$,

d_2 = the depth of the flanges in inches.

The Double Inverted T, Fig. 18.

b = the sum of the widths of the webs in inches,

b_2 = " " " " " flanges, or $MN - b$,

d_2 = the depth of the flanges in inches.

The ⊔ *Section,* Fig. 19.

b = the sum of the widths of the webs in inches,

b_2 = the width of the bottom in inches, or $OP - b$,

d_2 = the depth of the bottom in inches.

Noting the above definitions, the *moment* of *compressive resistance* for each section will be from Eq. 11, page 10,

$$R_c = \frac{b d_c^2 C}{6},$$
(60)

and for the *moment* of *tensile resistance* the formula given for the Hodgkinson beam will apply, or

$$R_\tau \text{ from Eq. 35, page 42.} \tag{61}$$

The Moment of Resistance of the section will be

$$R = R_c + R_\tau = 2R_c = 2R_\tau, \tag{62}$$

from the above equations.

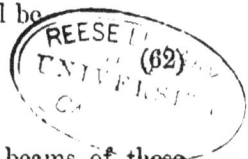

49. The Neutral Line. In cast-iron beams of these sections the neutral line will be given approximately by the following formula :

$$d_c = \frac{db + \sqrt{12bdb_2d_2(q+1) + b^2d^2(4q+5)}}{2b(q+1)}. \tag{63}$$

50. Transverse Strength. Place for R in Eq. 22 the values for the Moment of Resistance of the sections, and we will have

$$L = \frac{mbd_c^2C}{3s}, \tag{64}$$

and

$$L \text{ from Eq. 38, page 44.} \tag{65}$$

From either of these formulas the transverse breaking load may be computed.

51. To Design an Inverted T, Double Inverted T and ⊔ Sections.

Step I.—*Assume* a value for the depth, d, and the *width* or sum of the widths of the webs, b, and an economical position for the *neutral line*, and from these compute the other dimensions.

Step II.—The *compressed* area above the neutral line must have a moment of resistance equal to one half the Bending Moment of the applied load, L, or

$$\frac{bd_c^2C}{6} = \frac{Ls}{2m}. \tag{66}$$

If our *assumed* values for b and d_c do not satisfy this equation, other values must be assumed until the two members give identical numerical quantities.

FOR THE EXTENDED AREA :

Step III.—The load, L, sustained by the extended area of the web or webs, b, may be obtained by substituting their *moments* of *resistance* for R in Eq. 22, page 36, its value from Eq. 13, page 11,

$$L_1 = bd_\tau (2d + d_c) \frac{mT}{6s}. \qquad (67)$$

Step IV.—For the *flanges* and *bottom* deduct the load L_1 found by Step III., from one half of the applied load L, the remainder, L_2, is the load that must be sustained by the flanges and bottom.

$$L_2 = \frac{L}{2} - L_1. \qquad (68)$$

In order to *design* an area that will sustain this portion L_2 of the applied L, assume a convenient depth, d_2, and from this compute the width b_2, by substituting for R in Eq. 22, page 36, its value, the moment of tensile resistance for this case from Eq. 14, page 12,

$$L_2 = \left[3d_2 d_\tau (2d - d_2) - d_2^2 (3d - 2d_2) \right] \frac{mb_2 T}{6d_\tau s}, \qquad (69)$$

from which

$$b_2 = \frac{6d_\tau L_2 s}{3d_2 d_\tau (2d - d_2) - d_2^2 (3d - 2d_2) mT}, \qquad (70)$$

which gives the required width.

Section VI.—*Transverse Strength—Circular Sections.*

52. Moment of **Resistance.** In Fig. 20, let $BnOl$ represent the section, $BnOMnNB$ a section through the *strain wedges* on the line BO, B the origin of co-ordinates, DBS the *axis* or origin of moments, nl the *neutral line*, nBl the *compressed* area, and nOl the *extended* area

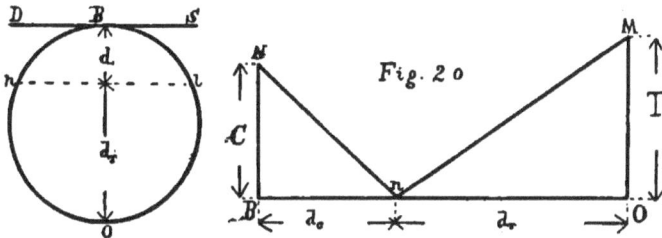

Fig. 20

Adopting the *notation* heretofore used and defined, we have

$C =$ the greatest intensity of the compressive strain in pounds, per square inch,

$T =$ the greatest intensity of the tensile strain in pounds, per square inch,

$R_\tau =$ the moment of the tensile resistance in inch-pounds,

$R_c =$ the moment of the compressive resistance in inch-pounds,

$R =$ the Moment of Resistance of the section,

$d_c =$ the versine Bn of one half of the arc nBl in inches,

$d_\tau =$ the versine On of one half of the arc nOl in inches,

$d = d_c + d_\tau =$ the diameter of the circle in inches,

$r =$ the radius of the circle in inches,

$q =$ the quotient arising from dividing C by T,

L = the total applied load in pounds,

s = the span, the distance between the supports in inches,

m = the factor defined in Art. 34, page 35,

Comp. arc = the arc bounding the compressed area in inches,

Tension arc = the arc bounding the extended area in inches.

For the moment of *compressive resistance* from Eq. 15, page 14, we have

$$R_c = \frac{C}{24d_c}\left[\sqrt{d_c d_{\tau}}\,[4d_c^{\,2}\,(d_c - r) + r^2\,(30r - 14d_c)] + Comp.\ arc\ (12d_c - 15r)r^2 \right], \quad (71)$$

and for the *moment* of *tensile resistance*

$$R_{\tau} = \frac{T}{24d_{\tau}}\left[\sqrt{d_c d_{\tau}}\,[4d_{\tau}^{\,2}\,(5r - d_{\tau}) + 18r^2\,(r - d_{\tau})] + Tension\ arc\ (12d_{\tau} - 9r)\,r^2 \right], \quad (72)$$

and for the Moment of Resistance of the circular section,

$$R = R_c + R_{\tau} = 2R_c = 2R_{\tau}. \quad (73)$$

53. The Neutral Line.   The position of the neutral line in circular wooden, cast-iron, wrought-iron, and steel beams will be found tabulated on pages 80, 100 and 112.

54. Transverse Strength.   Substituting for R, in Eq. 22, page 36, its value $2R_c = 2R_{\tau}$ from Eqs. 71 and 72, we have

$$L = \frac{mC}{12d_c s}\left[\sqrt{d_c d_{\tau}}\,[4d_c^{\,2}\,(d_c - r) + r^2\,(30r - 14d_c)] + Comp.\ arc\ (12d_c - 15r)\,r^2 \right], \quad (74)$$

and

$$L = \frac{mT}{12d_r^3} \left[\sqrt{d_c d_r} \left[4d_r^2 (5r - d_r) + 18r^2 (r - d_r) \right] \right.$$
$$\left. + \text{ Tension arc } (12d_r - 9r) r^2 \right], \quad (75)$$

from which the breaking load of a circular beam may be computed. By observing the following equalities, much time and labor will be economized in comparing the results computed from these formulas :

$$\sqrt{d_c d_r} \left[4d_c^2 (d_c - r) + r^2 (30r - 14d_c) \right]$$
$$= \sqrt{d_c d_r} \left[4d_r^2 (5r - d_r) + 18r^2 (r - d_r) \right],$$

$$\mp (12d_c - 15r) r^2 = \pm (12d_r - 9r) r^2,$$

$$Comp. \; arc = 2\pi r - \text{ Tension arc.}$$

Another Method. The Moments of Resistance of circular sections are to each other as the *cubes* of their *radii ;* hence by computing and tabulating the Moments of Resistance for all required positions of the neutral line, in a circle whose radius is *unity*, those for any other circle composed of the same material may be computed by multiplying the tabular number by the cube of the radius.

Let f_c = the second member of Eq. 74, when $r = 1$, except the factor $\frac{mC}{s}$,

f_r = the second member of Eq. 75, when $r = 1$, except the factor $\frac{mT}{s}$.

$$\therefore \; L = \frac{mr^3 f_c C}{s}, \quad (76)$$

and

$$L = \frac{mr^3 f_r T}{s}. \quad (77)$$

The application of Eq. 76 is illustrated in Examples 11, 32 and 33, and Eq. 77 in Examples 12, 13 and 14.

Tables giving the computed values of f_c and f_T, for the different positions of the neutral line in cast-iron, wrought-iron and steel, circular sections, whose radius is unity, are given on pages 80 and 100.

55. To Compute the Compressive Strain.

PROBLEM.—*The position of the neutral line and the compressive strain of the material of a circular section may be computed from the known transverse breaking load and tensile strength of the material.*

Deducing the value of f_T from Eq. 77, we have

$$f_T = \frac{Ls}{mr^3 T},$$ (78)

from the proper table in the sequel take the value of q corresponding to this value of f_T; then

$$C = qT.$$ (79)

Illustrated by Examples 3 and 23.

56. To Compute the Tensile Strain.

PROBLEM.—*The tensile strain of the material of a circular section may be computed from the known transverse breaking load and compressive strength of the material.*

From Eq. 76 deduce the value of f_c,

$$\therefore f_c = \frac{Ls}{mr^3 C}.$$ (80)

From the Table given in the sequel for the material take the value of q, corresponding to the computed value of f_c, then

$$T = \frac{C}{q}.$$ (81)

Illustrated by Example 24.

57. Relative Strength of Circular and Square Beams. Constant ratios exist between the Moments of Resistance of the circle and its inscribed, circumscribed and equal area square, when the material composing the square and the circle has the same tensile and compressive strength, and consequently the same ratios exist between the transverse strength of the beams of which they are sections, the span and manner of loading the circular and square beam being the same.

General Formula:

Let $f =$ the position of the neutral line in the square when the side is unity, and $d_c^2 = f^2d^2$, $b = d$ and $r' = \dfrac{d^3}{8}$, hence,

Strength of the Square $= \dfrac{bd_c^2 C}{3} = \dfrac{f^2d^3 C}{3}$, from Eq. 23.

Strength of the Circle $= r'f_c C = \dfrac{d^3f_c C}{8}$, from Eq. 76,

from which we have

$$Square : Circle :: \dfrac{d^3f^2 C}{3} : \dfrac{d^3f_c C}{8},$$

$$\therefore Circle = Square \times \dfrac{3d^3f_c}{8d^3f^2}. \tag{82}$$

CASE I.— *When the circle is inscribed within the square,* $d = d$, and Eq. 82 becomes

$$Circle = Square \times \dfrac{3f_c}{8f^2}. \tag{83}$$

Illustrated by Examples 15 and 16 of the sequel.

CASE II.— *When the square is inscribed within the circle side* of the *square,* $d = diameter$ of the *circle* $d \times 0.707$, and Eq. 82 becomes

$$Circle = Square \times \dfrac{3f_c}{2.8271 f^2}. \tag{84}$$

CASE III.— *When the square and circle are equal in area side* of the *square, d = diameter* of the *circle d* × 0.886, and Eq. 82 becomes

$$Circle = Square \times \frac{3\,f_c}{5.564\,f'^2}. \tag{85}$$

Illustrated by Examples 18 and 19.

———

SECTION VII.—*Transverse Strength—Hollow Circular Sections.*

58. Moment of Resistance. In Fig. 21, let *BnOl* represent the section, *BnOMnNB* a section through the *strain wedges* on the line *BO*, *B* the origin of co-ordinates, *DBS* the axis or origin of moments, and *nl* the neutral line.

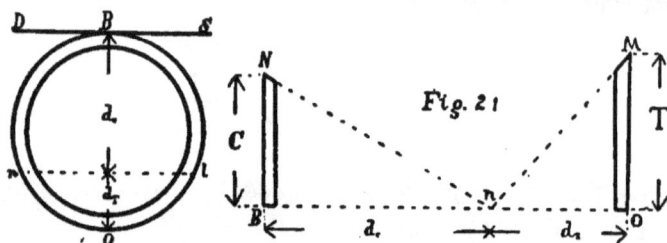

Fig. 21

Let *t* = the thickness of the metal ring in inches,
　　r = the radius of the outer circumference,
$r_1 = r - t =$ "　"　"　"　inner　"
　　tC = C and *tT = T.*

For other *notation* refer to Art. 52.

When *t*, the thickness of the metal ring, is *very small*, the entire strain distributed over the metal of the section may be treated as if it was all concentrated in the outer surface of the cylindrical beam; if t is not *very thin*, $r - \dfrac{t}{2}$ or the mean radius must be used instead of r in the following formulas.

Making the substitution of tC for C and tT for T in Eqs. 18 and 19, the *moments* of *compressive* and *tensile resistance* become

$$R_c = \frac{rtC}{2d_c}\left[\sqrt{d_c d_r} \times 2\,(3r - d_c) + Comp.\,arc\,(2d_c - 3r) \right],(86)$$

and

$$R_r = \frac{rtC}{2d_r}\left[\sqrt{d_c d_r} \times 2\,(r + d_r) + Tension\,arc\,(2d_r - r) \right].(87)$$

For the Moment of Resistance of the section,

$$R = R_c + R_r = 2R_c = 2R_r. \qquad (88)$$

59. The Neutral Line. The position of the neutral line in hollow circular sections of cast-iron, wrought-iron and steel, when the radius of the outer circumference is unity, will be found tabulated in the sequel.

60. Transverse Strength. Substituting for the *Moment* of *Resistance R*, in Eq. 22, page 36, its values in this case, $2R_c = 2R_r$, from Eqs. 86 and 87, we have

$$L = \frac{mrtC}{s\,d_c}\left[\sqrt{d_c d_r} \times 2\,(3r - d_c) + Comp.\,arc\,(2d_c - 3r) \right],(89)$$

and

$$L = \frac{mrtT}{s\,d_r}\left[\sqrt{d_c d_r} \times 2\,(r + d_r) + Tension\,arc\,(2d_r - r) \right].(90)$$

From either of the above formulas the transverse strength of hollow cylindrical beams may be computed.

In these formulas the following equalities exist:

$$\sqrt{d_c d_\tau} \times 2\,(3r - d_c) = \sqrt{d_c d_\tau} \times 2\,(r + d_\tau),$$

$$\pm\,(2d_c - 3r) = \mp\,(2d_\tau - r),$$

$$Comp.\ arc = 2\pi r - Tension\ arc.$$

When the metal ring is very thin, r in the above formulas is the radius of the outer circle, otherwise it is the radius of the mean circle, $r - \dfrac{t}{2}$.

Another Method. The Moments of Resistance of thin hollow circular sections are to each other as the squares of their radii; hence, by computing and tabulating the Moments of Resistance for all required positions of the neutral line, in a thin hollow circle whose radius is unity, those for any other thin hollow circle composed of the same material may be computed by multiplying the tabular number by the square of the radius.

Let f_c = the second member of Eq. 89, when radius = 1, except the factor $\dfrac{mtC}{8}$,

f_τ = the second member of Eq. 90, when radius = 1, except the factor $\dfrac{mtT}{8}$.

With this notation Eqs. 89 and 90 become

$$L = \frac{mr^2 f_c t C}{8}, \qquad (91)$$

and

$$L = \frac{mr^2 f_\tau t T}{8}. \qquad (92)$$

Tables giving the computed values of f_c and f_t for the positions of the neutral line in thin hollow circles, whose radius is *unity*, will be found in the sequel.

The application of the above formulas is illustrated in Examples 20 and 34.

CHAPTER IV.

SECTION I.—*General Conditions.*

61. Compressive Strength. *Crushing.*—The crushing strength of cast-iron that is usually obtained by experimenters and recorded for use in designing structures, is the number of pounds avoirdupois that it requires to crush a prism of the material whose sectional area is one square inch and length from one and a half to three times the diameter, under which condition it is found to be more nearly constant in value for the same material than when the height bears a greater or less ratio to the least diameter of the prism tested.

Value.—The range of values for the crushing strength of cast-iron may be taken as being from 85000 pounds to 125000 pounds per square inch; the mean is about 100000 pounds.

Elastic Limit.—The compressive elasticity of cast-iron as recorded by the earlier experimenters appears to be very defective, but improved methods of manufacture have produced a cast-iron from which modern experimenters find the increase in the amount of the compression to be, practically, in direct proportion to the increase in the load, within the elastic limit of the cast-iron.

The compressive elastic limit varies from three fifths to nearly the crushing strength.

62. Tensile Strength. *Tenacity* is the force in pounds that is required to pull asunder a prism of cast-iron whose sectional area is one square inch. It ranges in value from 15000 to 30000 pounds.

Elastic Limit.—The extensions within the elastic limit follow laws similar to those that govern the compressions. The elastic limit is about *one half* of the tensile strength.

63. Ratio of the Compressive to the Tensile Strength.

Experiments have demonstrated that the ratio existing between the *crushing* and *tensile* strength of cast-iron has a wider range of values than that for any other known material, and that this results from great fluctuations in the crushing rather than in the tensile strength. The extreme values for the $C \div T = q$ of our formulas may be taken at from 3 to $8\frac{1}{2}$; the tendency of improved methods of manufacture is to decrease, numerically, this ratio by increasing that of T and decreasing that of C, which in pure or wrought-iron becomes $C \div T = 1$.

64. Transverse Strength.

Cast-iron breaking with a well-defined fracture, its transverse strength may be computed from our formulas when the *crushing* and *tensile* strength of the *identical* cast-iron composing the beam is known from experiment, as the values of C and T are found to vary in an uncertain manner, with remelting, length of time in fusion, etc.

The *transverse strength* of some cast-iron does not increase directly with the increase in the dimensions of its cross-section, as it should in accordance with well-defined laws, but in a lower ratio. This defect is usually imputed to unequal strains being brought upon the metal in different parts of the section, from unequal temperatures in cooling, but few experiments have been made to test the matter.

Col. James, of England, planed out and tested a $\frac{3}{4}$-inch square bar from a 2-inch square cast bar, the values of C and T for this brand of iron, Clyde No. 3, having been determined from experiments by Mr. Eaton Hodgkinson. The following is the result of Col. James's tests :

Clyde No. 3.   The Dressed Bar.   Transverse Load.

$C = 106039$ lbs. $C = 60233$ lbs. by test.   193 lbs. rough bar.

$T = 23468$ " $T = 14509$ " by computation.*   134 " planed out.

From the above it appears that the value of C decreased in the "planed-out" bar 56.8 per cent; that of T 60.5 per cent, and the transverse breaking load 69.4 per cent. But this result is contrary to that obtained in the following tests made for the United States Government—Report of Tests of Iron and Steel for 1885.† In the United States Government tests the bars were cast two feet in length and three inches in width, with depths ranging from a half to two inches. The equivalent centre breaking load for a bar of the metal one inch square and twelve inches span was computed from the result of each experiment, that they might be compared.

Condition.	Rough.	Edges Planed.	Edges Planed.	Planed all over.
No. Expts.	4	3	6	6
Size,	$3'' \times 0''.5$	$2'' \times 1''$ to $1''.28$	$2'' \times 1''.5$ to $2''$	$2'' \times 1''.25$ to $1''.75$
Load, lbs.,	2453	2136	2025	2134

The "edges planed" bars were reduced in width from the rough castings that were 3″ wide, the tension and compression surfaces being left as they were cast.

The "planed-all-over" bars were cut from cast bars that were three inches wide and two inches deep, equal depths of metal having been removed from each face of the rough bar.

In these experiments the strength of the bars that were cast two inches deep varied about five per cent from those cast one inch deep, and the "planed-out" bars were as strong as the rough cast bars of the same size.

The bars that were tested with the tension and compression surfaces as cast varied in strength about 5 per cent from the average strength given above, and the "planed-out" bars

---

* Example No. 4.

† Senate Ex. Doc. No. 36—49th Congress, 1st Session, p. 1162.

only 2⅓ per cent in strength from the averages given, indicating great uniformity in strength, and consequently no decrease in the values of C and T for the metal in the interior of the cast bar, but this was not determined by direct experiment.

The following ratios are usually quoted to illustrate the depreciation of the transverse strength of cast-iron beams resulting from an increase in the size of the cross-section, the strength that the large beam should have being computed from that of the $1'' \times 1''$ bars, the latter being denoted by 100.

Experimenter.	$1'' \times 1''$	$2'' \times 2''$	$3'' \times 3''$
Capt. James......	100	71.84	61.95 per cent,
Mr. Hodgkinson..	100	71.22	"

The *transverse elastic limit load* may be computed when the corresponding values of C and T have been determined by experiment.

65. To Compute the Compressive and Tensile Strength. The relation existing between the Transverse Load, Compressive and Tensile Strength is such that the value of any one of the three may be computed when the value of the other two has been determined by direct experiment; this relation is true for both the *elastic limit* and the *breaking* values.

The beam from which the experimental breaking load was obtained must have been sufficiently long to deflect a distance equal to or greater than the depth of tension area that is required for *true* cross-breaking, or else our computed value of C or T will be incorrect for the reason given in Art. 24.

Compressive Strength.—This may be computed from the known tensile strength of the material and the transverse breaking load of rectangular and circular beams, or of that of any of the flanged beams when the neutral line lies within the lower or tension flange.

The cast-iron that Mr. Barlow used in his experiments was composed of pig and scrap-iron ; he determined its tensile and transverse strength, but he does not record in his " Strength of Materials " its crushing strength ; this we can compute from the data given.

EXAMPLE 1.—Required the Crushing Strength of Mr. Barlow's iron from the centre-breaking transverse load of a rectangular beam when

The depth..$d = 1$ inch, $T = 18750$ pounds by experiment,
" breadth $b = 1.02$ inch, $L =$ 534   "   "   "   *
" span...$s = 60.$ inches, $m =$ 4.

The position of the neutral line must be computed from Eq. 30, page 40,

$$d_1 = \frac{3 \times 1}{2} - \sqrt{\frac{9 \times 4 \times 1.02 \,(1)^2 \; 18750 - 12 \times 534 \times 60}{4 \times 4 \times 1.02 \times 18750}} = 0''.5032,$$

hence $d_c = 1 - 0.5032 = 0''.4968,$

from which we can compute the value of the crushing strain by means of Eq. 31,

$$C = \frac{3 \times 534 \times 60}{4 \times 1.02 \,(0.4968)^2} = 95462 \text{ pounds.}$$

The following Example is taken from Major Wade's Experiments on the " Strength and other Properties of Metals for Cannon," made for the United States Government :

EXAMPLE 2.—Required the Crushing Strength of Major Wade's cast-iron in third fusion, from the transverse strength of a rectangular beam, when

The depth...$d = 2.01$ inches, $T = 26569$ pounds by test,
" breadth $b = 2.008$ " . $L = 16172$ "   "   "
" span....$s = 20.$ " $m = 4$ for a centre load.

---

* Barlow's " Strength of Materials," p. 152.

For the position of the neutral line, we have from Eq. 30,

$$d_\tau = \frac{3 \times 2.01}{2} - \sqrt{\frac{9 \times 4 \times 2.008 \,(2.01)^2 \, 26569 - 12 \times 16172 \times 20}{4 \times 4 \times 2.008 \times 26569}} = 0''.8835,$$

$$\therefore d_c = 2.01 - 0.8835 = 1.1265 \text{ inches.}$$

Then we can compute the crushing strength, C, from Eq. 31,

$$C = \frac{3 \times 16172 \times 20}{4 \times 2.008 \,(1.1265)^2} = 95200 \text{ pounds.}$$

EXAMPLE 3.—Required the Crushing Strength of certain cast-iron from the transverse strength of a circular beam when

The diameter $d = 1''.129$, $T = 29400$ pounds mean of tests,
" span.....$s = 20''.0$, $L = 2118$ " by test,
$m = 4$.

From Eq. 78 we have

$$f_\tau = \frac{2118 \times 20}{4 \,(0.5645)^3 \, 29400} = 2.0024.$$

From the Table, page 80, for the above value of f_τ, we have $q = 3.275$, and from Eq. 79,

$$\therefore C = 29400 \times 3.275 = 96285 \text{ pounds.}$$

From the United States Government Report of the Tests of Iron and Steel for 1884,* the mean crushing strength of this cast-iron was $C = 100700$ pounds.

The *Crushing Strength* may be computed from the Transverse Strength of the Hodgkinson or any other flanged beam. When the neutral line is not situated above the top line of the tension flange, its position may be computed from Eq. 30, page 40, as if the beam were rectangular. When the neutral

---

* Senate Ex. Doc. No. 35—49th Congress, 1st Session, p. 284.

line is located above the top line of the tension flange, its position can be determined by the method given in Problem I., page 43, but in either case the value of C must be computed by formula 37, page 44, C being then the only unknown quantity that it will contain.

Tensile Strength.—This may be computed from the known Crushing Strength of the material and the Transverse Strength of any rectangular, circular or flanged beam.

EXAMPLE 4.—Required the Tensile Strength from the Transverse Strength of a rectangular beam made of a certain kind of cast-iron that was tested by Captain James, of England,[*] when

The depth $d = 0.75$ inches, $C = 60233$ pounds by test,
" breadth $b = 0.75$ " $L = 134$ "   "   "
" span $s = 54.0$ " $m = 4$

The position of the neutral line is to be computed from Eq. 32, page 40,

$$d_c = \sqrt{\frac{3 \times 134 \times 54}{4 \times 0.75 \times 60233}} = 0.4 \text{ inches,}$$

hence $d_x = 0.75 - 0.4 = 0.35$ inches.

Then we compute the required value of the tensile strength, T, from Eq. 33, page 40,

$$T = \frac{3 \times 134 \times 54}{4 \times 0.75 \times 0.35 \,(2 \times 0.75 + 0.4)} = 14509 \text{ pounds.}$$

Mr. Hodgkinson determined for this iron,

    Clyde, No. 3, $C = 106039$ and $T = 23468$ pounds.

---

* British "Report on the Application of Iron to Railway Structures," p. 257.

SECTION II.—*Rectangular Cast-Iron Beams.*

66. The Neutral Line. The position of the neutral line in rectangular cast-iron beams, for the different required ratios of $C \div T = q$, have been computed from Eq. 26, page 38, and tabulated below for reference.

TABLE *of positions of the neutral line in rectangular cast-iron beams.*

Ratio of Crushing to Tenacity or $C \div T = q$.	Depth of Neutral Line Below the Crushed Side of the Beam, or d_e.	Ratio of Crushing to Tenacity, or $C \div T = q$.	Depth of Neutral Line Below the Crushed Side of the Beam, or d_e.
8.0	0.4191 d	5.5	0.4831 d
7.875	0.4127 d	5.375	0.4871 d
7.75	0.4243 d	5.25	0.4913 d
7.625	0.4226 d	5.125	0.4956 d
7.5	0.4298 d	5.0	0.5000 d
7.375	0.4326 d	4.875	0.5045 d
7.25	0.4354 d	4.75	0.5092 d
7.125	0.4384 d	4.625	0.5139 d
7.0	0.4414 d	4.5	0.5189 d
6.875	0.4444 d	4.375	0.5240 d
6.75	0.4475 d	4.25	0.5293 d
6.625	0.4507 d	4.125	0.5347 d
6.5	0.4540 d	4.0	0.5403 d
6.375	0.4573 d	3.875	0.5461 d
6.25	0.4608 d	3.75	0.5521 d
6.125	0.4642 d	3.625	0.5583 d
6.0	0.4678 d	3.5	0.5647 d
5.875	0.4715 d	3.375	0.5714 d
5.75	0.4754 d	3.25	0.5784 d
5.625	0.4791 d	3.125	0.5855 d
		3.0	0.5930 d

Should lower numerical values of q be required, refer to the table of neutral lines in rectangular Sections of Steel.

67. Transverse Strength. From either Eq. 27 or 28, as may be most convenient, the transverse strength of rectangular cast-iron beams may be computed.

EXAMPLE 5.—Required the centre breaking load of a rect-angular cast-iron beam when

The depth........$d = 1$ inch, $C = 102434$ pounds by test.
"   breadth.......$b = 1$ " $T = 16724$ "   "   "
"   span.........$s = 54$ inches, $q = 6.125$ and $m = 4$.

The position of the neutral line may be computed from Eq. 26,

$$d_c = \frac{1(-0.5 + \sqrt{2 \times 6.125 + 2.25})}{6.125 + 1} = 0.4642 \text{ inches,}$$

or its value may be taken from the Table.   •
The transverse breaking load, L, from Eq. 28, becomes

$$L = \frac{4 \times 1 (0.4642)^2 \times 102434}{3 \times 54} = 545 \text{ pounds.}$$

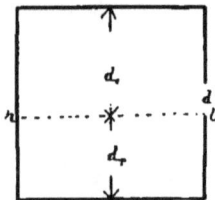

Mr. Hodgkinson, in his experiments, gives the above values of C and T for Blaen-Avon No. 2 iron, and 447 pounds as the mean transverse strength of 4 beams tested, while Captain James found 556 pounds to be the mean of 3 beams of the above dimensions for this brand of iron.

EXAMPLE 6.—Required the centre breaking load of the following: 1-inch square cast-iron beams, the span being 54 inches, and the values of C, T and the experimental breaking loads being from Mr. Hodgkinson's experiments. It is very probable that these bars did not deflect sufficiently to develop the true transverse strength.*

---

* Barlow's "Strength of Materials," p. 163.

	Tested Strength per Square Inch.		Neutral Line, or d_e.	Transverse Strength.	
	Crushing C.	Tensile T.		Computed.	Tested.
				lbs.	lbs.
Carron No. 2, C B....	106375	16683	0.457	548	476
" " 2, H. B...	108540	13505	0.418	468	463
" " 3, C. B....	115442	14200	0.416	493.	446
" " 3, II. B...	133440	17755	0.427	606	527
Buffery " 1, C. B....	93366	17466	0.488	549	463
" " 1, H. B...	86397	18434	0.456	429	436
Mean..				515	468

EXAMPLE 7.—Required the centre breaking load of a rectangular cast-iron beam, from Mr. Barlow's experiments, when

The depth.. $d =$ 2.00 inches, $C =$ 95462 lbs., computed, Ex. 1,
" breadth $b =$ 1.99 " $T =$ 18750 " by test,
" span....$s =$ 60.00 " $q =$ 5.091 and $m =$ 4.

For the neutral line we have, from Eq. 26, $d_c = 0''.9934$,

Computed transverse strength from Eq. 27 or 28 ....4175 lbs.,
Tested " " .... 3863 "

68. To Design a Rectangular Cast-Iron Beam.
The principles and formulas required are given in Art. 38, page 39.

SECTION III.—*Hodgkinson Cast-Iron Beams.*

69. In this section we will consider the principles as if applied to that class of cast-iron beams whose bottom or tension flange is larger than its top or compressed flange, though they are applicable to all other forms of flanged beams.

70. Mr. Hodgkinson's Experiments. Previous to the year 1840, when Mr. Hodgkinson commenced his experimental investigation into the strength of materials, the only metal beam that had been used in structures to any extent was the Inverted T, Fig. 17, and the Double T, Fig. 16, in which the top and bottom flanges were equal in area. Commencing his experiments with the equal flanged Double T, he found that, by increasing the area of the lower or tension flange by small amounts, he continued to increase the transverse strength of the beam, his unit of measure and standard of comparison being the quotient obtained from dividing the breaking load by the area of the section of the beam expressed in inches, and that this continued to increase until he had reached the point where the tension flange was about *six* times the area of the compressed flange; increasing it to a greater ratio he found that the transverse strength per square inch of section began to decrease. Experimenting with cast-iron, in which the ratio of C to T, or the crushing to the tensile strength, was about *six*, he recommended that the tension flange should have six times the area of the compressed flange, in order to obtain the greatest transverse strength per square inch of section. But subsequent investigators, experimenting with cast-iron possessing greater *tenacity*, or a less ratio of C to T than that used by Mr. Hodgkinson, found the greatest strength with a lower ratio of extended to compressed flange than that recommended by him. The reason for this will be apparent when the formulas for the strength of these beams are examined.

71. Neutral Line. When the neutral line is not situated within the *tension* flanges, its position may be *approximately* computed from Eq. 36, page 42. Should the neutral line lie within the *tension* flange, its position may be *accurately* computed from Eq. 30, or from the methods given in Probs. 1 and 2, page 43, when the transverse load and either the compressive or tensile strength has been determined from experiments.

72. Transverse Strength. The transverse strength of the Hodgkinson or of *any* beam having flanges at the top and the bottom may be computed from either Eq. 37 or 38, as may be most convenient.

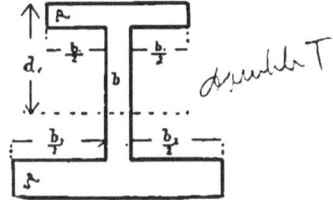

EXAMPLE 8.—Required the centre breaking load of a Hodgkinson cast-iron beam when

The depth of the beam......$d = 5.125$ ins., $C = 96000$ lbs.,
" breadth of the web......$b = 0.34$ " $T = 16000$ "
"   "   " top flanges, $b_1 = 1.22$ " $q = 6$
" depth   "   "   " $d_1 = 0.315$ " $s = 54$ ins.
"   "   " bottom " $d_2 = 0.56$ " $m = 4$.
" breadth   "   "   " $b_2 = 4.83$ "

Neutral Line.—Our formula 36, giving the position of the neutral line approximately only, places it in each case a little too near the tension flange. In this example it locates it within the tension flange, and we will *assume* that it coincides with its upper line, or

$$d_c = 4''.565 \text{ and } d_\tau = 0.56.$$

From Eq. 37 the transverse breaking load becomes

$$L = \frac{0.34 (4.565)^3 + 1.22 (0.315)^2 (3 \times 4.565 - 2 \times 0.315)}{3 \times 4.565 \times 54} \cdot 4 \times 96000 = 17616 \text{ lbs.}$$

The tension area being continuous from the neutral line to the bottom of the beam, Eq. 28, page 38, can be used to compute the load from the tensile strength T, in which $b = b + b_2$ of this example.

$$L = \frac{4 \times 5.17 \times 0.56\ (2 \times 5.125 + 4.565)\ 16000}{3 \times 54} = 16944\ \text{lbs.}$$

These two values of L, not being identical in numerical value, our assumed position for the neutral line must have been a little too low down in the section for exact equilibrium.

The above described beam was one of the series of beams that was used by Mr. Hodgkinson to determine the "section of greatest strength." Its transverse breaking load was, by test, 16730 pounds.* The value for T that we use was also determined from experiment; the ratio was supposed to be $C \div T = 6$. In this series of beams the neutral line of *rupture* was continuously lowered in the section, from progressively increasing the area of the tension flange until it finally reached the upper surface of this flange.

EXAMPLE 9.—Required the centre breaking load of a Hodgkinson Cast-Iron Beam (Fig. 15), when

The depth of the beam ....$d = 14.0$ ins., $C = 75983$ lbs. by test,
" breadth of the web.............$b = 1.0$ " $T = 13815$ " " "
"   "   "   top flanges.......$d_1 = 1.0$ " $q = 5.5$,
" depth "   "   " ....$b_1 = 2.5$ " $s = 16$ feet,
"   "   "   bottom flanges....$d_2 = 1.75$ " $m = 4$.
" breadth "   "   " ....$b_2 = 11.00$ "

From Eq. 36 the position of the neutral line becomes

$$d_e = \frac{1 \times 14 + \sqrt{12 \times 14 \times 11 \times 1.75\ (5.5 + 1) + (1 \times 14)^2\ (4 \times 5.5 + 5)}}{2 \times 1\ (5.5 + 1)} = 13.5\ \text{ins.,}$$

which is within the tension flanges, *assuming* $d_e = 14 - 1.75 = 12.25$.

---

* Barlow's " Strength of Materials," p. 177.

From Eq. 37 the breaking load, L, becomes

$$L = \frac{1\,(12.25)^3 + 2.5\,(1)^2\,(3 \times 12.25 - 2 \times 1)}{3 \times 12.25 \times 12 \times 16} \times 4 \times 75983 = 82013 \text{ lbs.}$$

From the tensile strength, T, and Eq. 28, L becomes

$$L = \frac{4 \times 12 \times 1.75\,(2 \times 14 + 12.25)\,13815}{3 \times 12 \times 16} = 80109 \text{ lbs.}$$

From this near identity of values for L we conclude that our assumed position for the neutral line was very near the correct position.

Two of these beams were constructed of Calder No. 1 cast-iron, and broken with centre loads of 73920 and 76160 pounds respectively by Mr. Owens, Inspector of Metals for the British Government.* The values of C and T were determined by Mr. Hodgkinson for this brand of iron. The bottom *table* contained *six* times the area of the top *table*, as recommended by Mr. Hodgkinson.

These beams broke with the deflections $1''.87$ and $2''.0$ respectively, and the full strength of the section was thus developed by the transverse load.

The following statement gives the proportion of the load sustained by each *member* of the transverse section and the proportion of the area that it occupied :

	Load.	Area.	
Compressed Flanges............	2.3	7.0	per cent.
" Web....	47.7	34.3	"
Extended "	4.2	5.0	"
" Flanges..........	45.8	53.7	"
	100.0	100.0	

The web extends through the total depth of the beam.

---

* Box's "Strength of Materials," p. 202.

73. To Design a Hodgkinson Beam. The formula required in designing a Hodgkinson beam will be found in Art. 42, page 44.

SECTION IV.—*Double T and Box Cast-Iron Beams.*

74. The Neutral Line. The position of the neutral line may be approximately computed from Eq. 36, page 42, by giving the letters of the formula their definitions in Art. 44.

75. Transverse Strength. The Transverse Strength may be computed from Eqs. 37 and 38, giving the letters their definition in Art. 44.

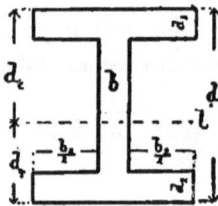

EXAMPLE 10.—Required the centre breaking load of a Hollow Rectangular Cast-Iron Beam whose outside dimension is $3''.125 \times 3''.125$, inside $2''.375 \times 2''.375$, thickness of metal all around 0.375, when

The depth of the beam......$d = 3''.125$ $C = 84000$ lbs.

" breadth " sides.....$b = 2 \times 0''.375$ $T = 14000$ "

" " " top........$b_1 = 2''.375$ $q = 6$,

" depth " " .....$d_1 = 0''.375$ $s = 6$ feet,

" " " bottom...$d_2 = 0''.375$ $m = 4$.

" breadth " " ...$b_2 = 2''.375$

The position of the neutral line d_c from Eq. 36 becomes

$$d_c = \frac{3.125 \times 0.75 + \sqrt{12 \times 3.125 \times 0.75 \times 2.375\,(6+1) + (3.125+0.75)^2\,(4 \times 6+5)}}{2 \times 0.75\,(6+1)} =$$

$$1''.965.$$

The breaking load from Eq. 37 becomes

$$L=\frac{0.75\,(1.965)^3+2.375\,(0.375)^3\,(3\times1.965-2\times0.375)}{3\times1.965\times12\times6}\times4\times84000=5760\ \text{lbs.},$$

and from the tensile strength T, and Eq. 38, L becomes

$$L = 5178 \text{ pounds.}$$

In the investigations preliminary to the construction of the Menai Straits Tubular Bridge, Mr. Stephenson broke this beam with 5387 pounds.*

76. To Design a Double T and Hollow Rectangular Cast-Iron Beam. This may be done by using the formula and directions given in Art. 47, page 48.

---

* "Britannia and Conway Tubular Bridges," p. 429.

Section V.—*Circular Cast-Iron Beams.*

77. Neutral Line. *Table* of *positions* of *the neutral line* in Circular Cast-Iron Beams and factors for use in Eqs. 76 and 77.

Ratio of Crushing to Tenacity, or $C \div T = q$.	Depth of Neutral Line Below the Crushed Side of the Beam, or d_c.	FACTORS FOR COMPUTING THE MOMENT OF RESISTANCE, $r = 1$.	
		f_c for Crushing Strain, C.	f_τ for Tensile Strain, T.
8.	0.3984 d	0.3226	2.5804
7.875	0.4004 d	0.3260	2.5722
7.75	0.4022 d	0.3298	2.5640
7.625	0.4041 d	0.3336	2.5556
7.5	0.4062 d	0.3380	2.5466
7.375	0.4087 d	0.3426	2.5356
7.25	0.4112. d	0.3472	2 5250
7.125	0.4137 d	0.3518	2 5138
7.0	0.4161 d	0.3568	2.5030
6.875	0.4186 d	0.3620	2.4920
6.75	0.4201 d	0.3672	2.4808
6.625	0.4236 d	0.3726	2.4694
6.5	0.4262 d	0.3780	2.4578
6.375	0.4289 d	0 3836	2.4458
6.25	0.4316 d	0.3892	2.4346
6.125	0.4344 d	0.3952	2.4208
6.0	0.4373 d	0.4014	2.4080
5.875	0.4402 d	0.4078	2.3952
5.75	0.4432 d	0.4142	2.3812
5.625	0.4463 d	0.4210	2.3664
5.5	0.4494 d	0.4278	2.3536
5.375	0.4526 d	0.4350	2.3392
5.25	0.4559 d	0.4426	2.3240
5.125	0.4594 d	0.4504	2.3080
5.0	0.4629 d	0.4584	2.2916
4.875	0.4665 d	0.4668	2.2750
4.75	0.4702 d	0.4752	2.2580
4.625	0.4740 d	0.4842	2.2400
4.5	0.4780 d	0.4938	2.2218
4.375	0.4821 d	0.5036	2.2028
4.25	0.4863 d	0.5138	2.1832
4.125	0.4906 d	0.5244	2.1630
4.0	0.4951 d -	0.5356	2.1424
3.886	0.5000 d	0.5478	2.1186
3.75	0.5045 d	0 5593	2.0974
3.625	0.5095 d	0.5722	2.0739
3.5	0.5146 d	0.5855	2.0493
3.375	0.5200 d	0.5996	2.0237
3.25	0.5256 d	0 6145	1.9970
3.125	0 5314 d	0.6302	1.9691
3.0	0.5375 d	0 6468	1.9399

From this Table the position of the neutral line is obtained by placing for d, the diameter in inches and computing the product indicated.

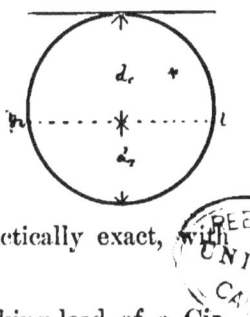

78. The Transverse Strength of Circular Cast-Iron Beams may be accurately computed from either Eq. 74 or 75, but as this involves a tedious calculation Eq. 76 or 77 will give results, practically exact, with much less labor.

EXAMPLE 11.—Required the centre breaking load of a Circular Cast-Iron Beam, when

The diameter $d = 2''.0$, $C = 95200$ pounds computed, Ex. 2,
" span......$s = 20''.0$, $T = 26569$ " by test,
$m = 4$, $q = 3.58$.

The values of the factors f_e and f_τ, obtained by interpolation from the Table, page 80, and substituted in Eqs. 76 and 77, give

$$L = \frac{4\,(1)^2\,0.577 \times 95200}{20} = 10986 \text{ pounds.}$$

Major Wade, in his experiments, broke this beam with 11112 pounds.

EXAMPLE 12.—Required the centre breaking load of a Circular Cast-Iron Beam when

The diameter...$d = 2''.42$, $C = 95200$ lbs. computed, Ex. 2,
" span ....$s = 20''.0$, $T = 26569$ " by test,
$m = 4$, $q = 3.58$.

$f_\tau = 2.065$ from the Table, which, substituted in Eq. 77, gives

$$L = \frac{4\,(1.21)^2\,2.065 \times 26569}{20} = 19439 \text{ pounds.}$$

Major Wade broke four beams of the above dimensions with 18141, 20419, 19997 and 18225 pounds respectively, the mean strength being 19198 pounds.

EXAMPLE 13.—Required the centre transverse *elastic limit* load of a Circular Cast-Iron Beam, when

The diameter......$d = 1''.129$, $C = 20000$ pounds by test,
"   span..........$s = 20''.0$, $T = 17000$ "   "
          $m = 4$, $q = 1.111$.

The value of $f_{\tau} = 1.1818$ from the Table, page 80, substituted in Eq. 77, gives

$$L = \frac{4 (0.5645)^{2} 1.1818 \times 17000}{20} = 1062 \text{ pounds.}$$

The *elastic limit* load, of the two of these beams that were tested, was 1130 pounds each, the values of C and T being the mean of three tests, as given in the United States Government Report of the Tests of Iron and Steel for 1884.*

EXAMPLE 14.—Required the centre breaking load of the Circular Cast-Iron Beam whose *elastic limit* load was computed in Example 13, when

$C = 100700$ pounds by test, $f_{\tau} = 2.0333$ from the Table,
$T = 29400$ "   " $q = 3.422$.

From Eq. 77 we have

$$L = \frac{4 (0.5645)^{2} 2.0333 \times 29400}{20} = 2131 \text{ pounds.}$$

The breaking load of the two beams described in Example 13 was 2118 and 1795 pounds respectively.

79. Movement of the Neutral Line. In the Circular Cast-Iron Beam, Example 13, the depth of the neutral

---

* Senate Ex. Doc. No. 35--49th Congress, 1st Session, p. 284.

line below the compressed surface at the *elastic limit* was, from the Table, page 80, $d_c = 0''.7853$ in Example 14, at the instant of *rupture*, $d_c = 0''.5846$; hence, as the loading progressed, the *neutral line moved upward*, or from the tension side toward the compressed side of the beam.

80. Relative Strength of Circular and Square Cast-Iron Beams.

CASE I.— *When the circle is inscribed within the square.* The relation required is given in Eq. 83.

EXAMPLE 15.—Required the relation between the transverse strength of a square and the *inscribed* circular cast-iron beam when $C \div T = 5$.

$f = 0.5$ from Table, page 71, $f_c = 0.4584$ from Table, page 80.

$$\frac{3 \times f_c}{8f^2} = \frac{3 \times 0.4584}{8 \times (0.5)^2} = 0.6876,$$

$\therefore$ *Strength of Circle = Strength of the Square* $\times$ 0.6876.

EXAMPLE 16.—Required the centre breaking transverse load of the circular beam inscribed within the square, from the tested strength of the square beam, when

The side of the square $d = 2''.01$, $q = 3.58$,
" diameter of the circle $d = 2''.01$, $f_c = 0.577$ from the Table,
" span . . . . $s = 20''.0$, $f = 0.561$ " " "

With these values Eq. 83 becomes

$$\frac{3 \times f_c}{8 \times f^2} = \frac{3 \times 0.577}{8\,(0.561)^2} = 0.687.$$

Actual b'k'g load square beam, Maj. Wade's tests, 16172 lbs.
" " " circular " " " " 11112 "
Comp'd " " " " $= 16172 \times 0.687 = 11110$ "

CASE II.— *When the square is inscribed within the circle.* The relation will be given by Eq. 84.

EXAMPLE 17.—Required the relation between the strength

of the circular and its inscribed square beam, when $C \div T = 5$. $f = 0.5$ from the Table, page 71, $f_c = 0.4584$ from the Table, page 80.

$$\therefore \frac{3f_c}{2.8271\,f^2} = \frac{3 \times 0.4584}{2.8271\,(0.5)^2} = 1.945,$$

$\therefore$ *Strength of the Circle = Strength of the Square* $\times$ 1.945.

CASE III.— *When the circular is equal in area to that of the square beam.*

The relation will be given by Eq. 85.

EXAMPLE 18.—Required the relation between the transverse strength of the circular and the square beam, when their areas are equal and $C \div T = 5$.

$f = 0.5$ from the Table, page 71, $f_c = 0.4584$ from the Table, page 80.

$$\therefore \frac{3f_c}{5.564\,f^2} = \frac{3 \times 0.4584}{5.564 \times (0.5)^2} = 0.987,$$

$\therefore$ *Strength of the Circle = Strength of the Square* $\times$ 0.987.

EXAMPLE 19.—Required the centre breaking load of a Circular Cast-Iron Beam from that of the square beam of equal area, when

The side of the square....$d = 1''.01$, $q = 5.091$,
" diameter of the circle $d = 1''.145$, $f_c = 0.4562$, from the Table, page 80,
" span.................$s = 60''.0$ $f = 0.496$ " " " " 71.

With these values Eq. 85 becomes

$$\frac{3 \times 0.4562}{5.564\,(0.496)^2} = 1.$$

Mean b'k'g l'd of 4 of these square beams from Mr. Barlow's tests,[*] 519 lbs.,
" " " of the circular beam from Mr. Barlow's tests, 519 pounds,
Comp'd " " " " " $= 519 \times 1 = 519$ pounds.

---

[*] Barlow's "Strength of Materials," p. 152.

SECTION VI.—*Hollow Circular Cast-Iron Beams.*
81. Neutral Line.

Ratio of Crushing to Tenacity, or. $C \div T = q$.	Depth of Neutral Line Below the Crushed Side of the Beam, or d_c.	FACTORS FOR COMPUTING THE MOMENT OF RESISTANCE, $r = 1$.	
		f_c for Crushing Strain, C.	f_τ for Tensile Strain, T.
8.319	0.5000 d	0.8584	7.1414
8.25	0.5020 d	0.8640	7.1282
8.125	0.5057 d	0.8742	7.1104
8.0	0.5093 d	0.8852	7.1820
7.875	0.5131 d	0.8962	7.0580
7.75	0.5170 d	0.9076	7.0352
7.625	0.5209 d	0.9190	7.0078
7.5	0.5249 d	0.9304	6.9818
7.375	0.5290 d	0.9430	6.9550
7.25	0.5331 d	0.9556	6.9278
7.125	0.5374 d	0.9684	6.8996
7.0	0.5418 d	0.9816	6.8704
6.875	0.5462 d	0.9950	6.8410
6.75	0.5507 d	1.0090	6.8104
6.625	0.5553 d	1.0232	6.7790
6.5	0.5600 d	1.0380	6.7470
6.375	0.5647 d	1.0534	6.7174
6.25	0.5697 d	1.0688	6.6800
6.125	0.5748 d	1.0848	6.6442
6.0	0.5800 d	1.1008	6.6092
5.875	0.5852 d	1.1184	6.5694
5 75	0.5907 d	1.1356	6.5296
5.625	0.5962 d	1.1540	6.4918
5.5	0.6018 d	1.1730	6.4492
5.375	0.6076 d	1.1924	6.4062
5.25	0.6134 d	1.2124	6.3640
5.125	0.6193 d	1.2330	6.3184
5.0	0.6255 d	1.2542	6.2702
4.875	0.6314 d	1.2748	6.2266
4.75	0.6384 d	1.2994	6.1718
4.625	0.6451 d	1.3230	6.1192
4.5	0.6520 d	1.3482	6.0638
4.375	0.6590 d	1.3740	6.0064
4.25	0.6662 d	1.4000	5.9492
4.125	0.6735 d	1.4274	5.8877
4.0	0.6811 d	1.4559	5.8240
3.875	0.6889 d	1.4856	5.7569
3.75	0.6970 d	1.5168	5.6890
3.625	0.7052 d	1.5488	5.6148
3.5	0.7137 d	1.5826	5.5387
3.375	0.7223 d	1.6172	5.4578
3.25	0.7313 d	1.6538	5.3745
3.125	0.7405 d	1.6922	5.2880
3.0	0.7500 d	1.7320	5.1961

The above Table gives the positions of the neutral line in Hollow Circular Cast-Iron Beams, and factors for use in Eqs. 91 and 92.

82. Transverse Strength.

The transverse strength of Hollow Cast-Iron Beams may be accurately computed from either Eq. 89 or 90, also from Eq. 91 or 92, and with much less labor.

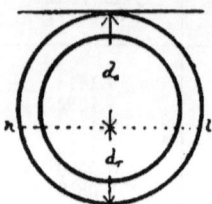

EXAMPLE 20.—Required the centre breaking load of a Hollow Circular Cast-Iron Beam, when

The outer diameter.... $= 3''.875$, $C = 84000$ pounds,
" inner " .... $= 3''.125$, $T, = 14000$ "
" mean " .. $d = 3''.5$, $s, = 6$ feet,
" thickness of metal $t = 0''.375$, $f_c, = 1.1008$ from the Table.

By using these values in Eq. 91 and for $r = 1.75$, one half of the mean diameter the load, L, becomes

$$L = \frac{4 \times 0.375 \ (1.75)^2 \ 1.1008 \times 84000}{12 \times 6} = 5899 \text{ pounds.}$$

The actual breaking load was 5122 pounds from the series of experiments described in Example 10.

CHAPTER V.

SECTION I.—*General Conditions— Wrought-Iron.*

83. Compressive Strength. *Crushing.*—Wrought-iron, when subjected to pressure, increases its area so rapidly that it is impossible to determine with precision its *crushing* strength or that intensity of pressure that corresponds to the *tenacity.* Its *value*, per square inch, as quoted by the various writers, varies from 30000 to 90000 pounds per square inch.

The great pressure to which wrought-iron is subjected when rolled into sheets and eye-beams causes it to lose a portion of its ductility and increase its crushing strength. The crushing strength of Swedish bar-iron, computed from Mr. Kirkaldy's experiments (Example 21), is 60000 pounds per square inch, while the crushing strength of wrought-iron when rolled into eye-beams, computed from Mr. Fairbairn's experiments, is 86466 pounds per square inch (Example 22).

Elastic Limit.—The compressive elastic limit of wrought-iron in bars may be taken at 30000 pounds per square inch. The computed value of the compressive elastic limit for rolled eye-beams is 63170 pounds (Example 31), from data furnished by the Phœnix Iron Company's experiments.

84. Tensile Strength. *Tenacity.*—When wrought-iron is subjected to a tensile strain before rupture takes place, it increases in length from 15 to 20 per cent, and contracts in area at the fractured section about 25 per cent. Its *tenacity* ranges in value from 50000 to 65000 pounds per square inch ; wrought-iron diminishes in tensile strength when rolled into sheets.

Elastic Limit.—The tensile elastic limit of bar-iron is about *one half* of its tenacity; the mean may be considered to be 30000 pounds per square inch.

STEEL.

85. Compressive Strength. *Crushing.*—In determining its crushing strength, the same difficulty mentioned in determining that of wrought-iron is encountered ; its value, as given for the different kinds, varies from 100000 to 340000 pounds per square inch.

The *compressive elastic limit* of steel ranges from 20000 to 60000 pounds per square inch.

86. Tensile Strength. The *tenacity* of steel is greater than that of any known material; it ranges in value from 60000 to 160000 pounds per square inch.

87. To Compute the Compressive Strength. Formula 31, page 40, and Eq. 79, page 58, may be used to great advantage to determine the crushing strength for such materials as wrought-iron and steel whose resistance to crushing cannot be readily determined in the usual manner, because they increase their area so rapidly, under direct pressure, when applied to small test specimens, that no precise determination of their strength can be made. The deflection of the beam, however, must be sufficient to allow the *neutral line* to move to the position required for rupture of the fibres by the compressive and tensile strains at the same instant, for the reason given in Art. 24.

EXAMPLE 21.—Required the Crushing Value of C for wrought-iron, from the tensile strength and the centre transverse *breaking-down* load, when

The depth.....$d = $ 2.0 inches, $T = 42133$ pounds, by test,
" breadth...$b = $ 2.0 " $L = 13338$ " " "
" span......$s = 25.0$ " $m = 4.$

The required position of the neutral line from Eq. 30 is

$$d_{\mathrm{r}} = \frac{3 \times 2}{2} - \sqrt{\frac{9 \times 4 \times 2 \,(2)^2\, 42133 - 12 \times 13338 \times 25}{4 \times 4 \times 2 \times 42133}} = 0.554,$$

$\therefore d_{\mathrm{c}} = 2 - 0.554 \doteq 1.446$ inches and C from Eq. 31.

$$C = \frac{3 \times 13338 \times 25}{4 \times 2 \,(1.446)^2} = 60090 \text{ pounds.}$$

This example is taken from Mr. Kirkaldy's experiments.[*] By direct .pressure he determined $C = 84890$ pounds, when the length $= 2$ diameters, and $C = 148840$ pounds, when the length $= 1$ diameter of the test specimen. Our computation of the value of C is made under the hypothesis that the tensile strain in the experiment was sufficiently *intense*, though it did not actually fracture the beam.

EXAMPLE 22.—Required the Crushing Value of C for rolled wrought-iron from the tensile strength and the centre transverse breaking load of a rolled wrought-iron T beam.

The depth of the beam $d = 3''.0$, $T = 57600$ pounds mean for British bar-iron.

The breadth of the web $b = 0''.5$, $L = 2690$ pounds, from Mr. Fairbairn's test.[†]

The breadth of the compressed flanges $b_1 = 2''.5$, $m = 4$.
The depth   "   "   " $d_1 = 0''.375$, $s = 10$ ft.

The position of the neutral line, the beams having no bottom flanges, from Eq. 30, becomes

$$d_{\mathrm{r}} = \frac{3 \times 3}{2} - \sqrt{\frac{9 \times 4 \times 0.5 \,(3)^2\, 57600 - 12 \times 2690 \times 120}{4 \times 4 \times 0.5 \times 57600}} = 1.06 \text{ ins.,}$$

---

[*] Barlow's "Strength of Materials," p. 256.
[†] Box's "Strength of Materials," p. 214.

and from Eq. 37,

$$L = 2690 = \frac{0.5\,(1.94)^3 + 2.5\,(0.375)^2\,(3 \times 194 - 2 \times 0.375)}{3 \times 1.94 \times 12 \times 10} = 0.03111\ C,$$

$$\therefore\ C = \frac{2690}{0.03111} = 86466 \text{ pounds.}$$

EXAMPLE 23.—Required the Crushing Strength of certain steel made by the Otis Iron and Steel Co., of Cleveland, Ohio, from the centre breaking load of a circular beam when

The diameter $= 1''.129$, $T = 83500$ pounds mean of 10 tests,
" span $\quad = 20''.0$, $L = 3860$ " one test.

From Eq. 78 we have

$$f_{\scriptscriptstyle T} = \frac{3860 \times 20}{4\,(0.5645)^3\,83500} = 1.2849.$$

From the Table, page 100, $q = 1.2122$ for this value of $f_{\scriptscriptstyle T}$, and from Eq. 79,

$$C = 1.2122 \times 83500 = 101218 \text{ pounds.}$$

This example is taken from the United States Government Report of Tests of Iron and Steel, at Watertown Arsenal, for 1885.*

88. To Compute the Tensile Strength. The tensile strength may be computed from test values of C and the load L in rectangular beams by means of Eq. 33, having first ascertained the position of the neutral line from Eq. 32, and for circular beams from Eqs. 80 and 81; page 58.

EXAMPLE 24.—Required the Tensile Strength of " Burden's Best " wrought-iron from the centre breaking load of a circular beam when

---

* Senate Ex. Doc. No. 36—49th Congress, 1st Session, p. 690.

The diameter.......$d = 1''.25$, $C = 64500$ pounds assumed,
" span...........$s = 12''.0$, $L = 6000$ " by test.

From Eq. 80 we have

$$f_c = \frac{6000 \times 12}{4\,(0.625)^2\,64500} = 1.15.$$

From the Table, page 100, $q = 1$ for the above value of f_c, from Eq. 81,

$$T = \frac{64500}{1} = 64500 \text{ pounds.}$$

This beam was broken at the Rensselaer Polytechnic Institute in November, 1882, as given by Professor W. H. Burr.

89. Transverse Strength of Wrought-Iron.
The elastic limit load is technically called the breaking load for wrought-iron beams, as from its position in the section of the beam where fracture should take place from the fibre strains being greatest, it cannot undergo the change of form required before its fibres can be *fractured ;* this breaking load may be computed from the tensile and compressive *elastic limit* coefficients of wrought-iron.

The breaking-down load.—While wrought-iron in beams cannot be broken transversely by rupturing its fibres, yet there is an intensity of the transverse load at which it does not appear to be able to offer any further resistance to the action of the bending load; this load may be called the *breaking-down* load; in wrought-iron beams it is about twice the *technical* breaking or elastic limit load.

Wrought-iron when rolled into T beams may become sufficiently hard and unyielding from the intense pressure required to make it fill the rolls, that it will offer sufficient resistance to actually fracture the fibres by the tensile strain, as in Example 22.

90. Transverse Strength of Steel. As in the case of wrought-iron beams the elastic limit load of steel beams is technically called the breaking load.

The *breaking-down* load of steel beams is about $\frac{4}{5}$ths of the elastic limit load.

SECTION II.—*Rectangular Wrought-Iron and Steel Beams.*

91. Neutral Line. The position of the neutral line may be computed from either formula 25 or 26. From Eq. 25 the position has been computed for the different values of $C \div T = q$, that are required in wrought-iron and steel rectangular beams, and the results tabulated below for reference.

Table of Positions of the Neutral Line in Rectangular Wrought-Iron and Steel Beams :

Ratio of Crushing to Tenacity, or $C \div T = q$.	Depth of Neutral Line Below the Crushed Side of the Beam, or d_c.	Ratio of Crushing to Tenacity, or $C \div T = q$.	Depth of Neutral Line Below the Crushed Side of the Beam, or d_c.
3.0	0.5930 d	1.9	0.6757 d
2.9	0.5993 d	1.8	0.6852 d
2.8	0.6057 d	1.7	0.6951 d
2.7	0.6124 d	1.6	0.7056 d
2.6	0.6193 d	1.5	0.7161 d
2.5	0.6264 d	1.4	0.7280 d
2.4	0.6338 d	1.3	0.7401 d
2.3	0.6416 d	1.2	0.7529 d
2.2	0.6496 d	1.1	0.7664 d
2.1	0.6579 d	1.0	0.7807 d
2.0	0.6666 d	0.9	0.7960 d

For ratios greater than $C \div T = 3$ refer to the Table, page 80.

92. Transverse Strength. Either Eq. 27 or 28 may be used to compute the transverse strength of rectangular wrought-iron and steel beams, as may be most convenient.

EXAMPLE 25.—Required the centre breaking or elastic limit load of a Rectangular Wrought-Iron Beam, when

The depth ....$d =$ 3".0, $C = $ 30000 pounds mean value,
" breadth .....$b = $ 1".5, $T = $ 30000 " " "
" span ....$s = $ 33".0, $q = 1$, and $m = 4$.

The position of the neutral line may be computed or taken directly from the Table for $q = 1$.

$d_c = 0.7805 \times 3 = 2".3415$, $d_\tau = 3 - 2.3415 = 0.6585$.

From Eq. 28 the value of the load, L, required becomes

$$L = \frac{4 \times 1.5 \times 0.6585\,(2 \times 3 + 2.3415)\,30000}{3 \times 33} = 9681 \text{ pounds.}$$

Mr. Barlow tested this iron and found its tensile elasticity *perfect* with 22400 pounds ; the transverse elastic limit load was, by experiment, between 9520 and 10080 pounds. A number of beams tested gave similar results.

The deflection with 10080 pounds was 0".963 in one bar and 0".624 in another; hence the elastic fibre strain limits were fully developed.*

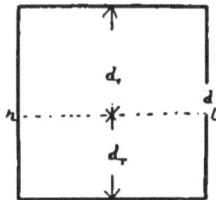

EXAMPLE 26.—Required the centre breaking or elastic limit load of a Rectangular Wrought-Iron Beam of Swedish iron, when

The depth..........$d = $ 2".0, $C \doteq 22637$ pounds by test,
" breadth........$b = $ 2".0, $T = 24052$ " "
" span..........$s = 25".0$, $q = 0.94$, and $m = 4$.

The position of the neutral line may be computed from Eq. 25, or taken from the Table, by proportion between the posi-

---

tions for $q = 1$ and $q = 0.9$, for which $d_c = 1.58$. With these values the required load, L, becomes from Eq. 27,

$$L = \frac{4 \times 2 (1.58)^2 \, 22637}{3 \times 25} = 6003 \text{ pounds.}$$

Mr. Kirkaldy determined, experimentally, that the centre breaking load of this beam was between 6000 and 6500 pounds, and that the *elastic limit* values of C and T were as given above.

The position of the neutral line for the *breaking-down* load of this beam (Example 21) was 1″.446 below the top or compressed side of the beam, and as the loading progressed it must have moved upward from its position at the elastic limit, 1″.58, to 1″.446, its position with the *breaking-down* load.

The deflection with 6000 pounds was 0″.378, and with 6500 0″.506 ; it only required 0″.46 to fully develop the elastic fibre strain limits.*

EXAMPLE 27.—Required the centre transverse elastic limit load of a Rectangular Steel Beam, when

The depth..........$d =$ 1″.75, $C = 48000$ pounds by test,
" breadth........$b =$ 1″.75, $C = 52000$ " "
" span...........$s =$ 25″.0, $q = 0.923$, and $m = 4$.

The position of the neutral line, computed from Eq. 25, or taken from the Table by proportion, is

$$d_c = 1″.3867 \text{ and } d_\tau = 0.3633.$$

The required load, L, from Eq. 27, becomes

$$L = \frac{4 \times 1.75 (1.3867)^2 \, 48000}{3 \times 25} = 86147 \text{ pounds.}$$

---

* Barlow's "Strength of Materials," p. 256.

Mr. Kirkaldy determined by experiment that the elastic limit load of each of 4 of these beams was between 8000 and 9000 pounds, and that the values of C and T were as given.

The deflection with 8000 pounds was $0''.476$, and it only required $0''.3633$ to develop the elastic compressive and tensile fibre strain limits at the same instant.*

EXAMPLE 28.—Required the *breaking-down* load of the beam described in Example 27, when

$$C = 159582 \text{ pounds by test, } q = 2.3,$$
$$T = 69336 \quad `` \quad `` \quad m = 4.$$

For the position of the neutral line from Eq. 25 or from the Table,

$$d_c = 1''.121 \text{ and } d_r = 0''.629.$$

The breaking-down load, L, becomes from Eq. 27,

$$L = \frac{4 \times 1.75 \, (1.121)^2 \, 159582}{3 \times 25} = 18716 \text{ pounds.}$$

The mean *breaking-down* load of 4 of these beams was 16477 pounds from Mr. Kirkaldy's experiments. The upward movement of the neutral line is seen in these Examples, in Example 27, $d_c = 1''.3867$ and in Example 28, $d_c = 1''.121$.

---

* Barlow's " Strength of Materials," p. 253.

SECTION III.—*Double T, Rolled Eye-Beams and Hollow Rectangle or Box, Wrought-Iron and Steel Beams.*

93. Neutral Line. The neutral lines in these wrought-iron and steel beams are so near the bottom or tension side of the beam that their positions cannot be computed from approximate formulas, and the *exact* formula is too complicated to be useful in practice. Before beams of these sections are manufactured, its position should be assumed and a sufficient area of the metal placed above and below the *assumed* neutral line to insure equilibrium from the known values of the crushing and tensile strength of the material.

In most practical examples of these beams the neutral line is located within the tension flange, and its position can be computed from Eq. 30, page 40, when *L* the load and *T* the tensile strength have been determined experimentally, or by the methods given in Problems I. and II., page 43.

94. Transverse Strength. When the position of the neutral line is known the transverse strength may be computed from either Eq. 37 or 38, that were deduced for the transverse strength of the Hodgkinson beam, giving the letters of the formulas the meaning defined in Art. 44, page 47.

The transverse strength of both the Tee and Double-Headed Railroad Rails may be very accurately computed from the above formulas, although they are partly bounded in outline by curved lines. The web must extend from the bottom to the top of the rail, and an equivalent in area right line section must be formed for the metal that remains in both the base and head of the rail, which will enable the formulas to be applied to these forms of beams.

95. Double T *beams with tension and compression flanges unequal in area.*

EXAMPLE 29.—Required the centre elastic limit load of a Wrought-Iron Rolled Double T Beam, when

Depth of beam....... $d = 8''.38$, $C = 36320$ lbs. required for eq'brium,
Breadth of web........$b = 0''.325$, $T = 30000$ pounds mean value,'
   "   " top flanges $b_1 = 2''.175$, $m = 4$,
Depth   "   "   " $d_1 = 1''.0$, $s = 11$ feet,
   "   " bottom " $d_2 = 0''.38$,
Breadth "   "   " $b_2 = 3''.675$.

The neutral line will be assumed to be near the top line of the tension flanges, or

$$d_c = 8''.05 \text{ and } d_\tau = 0.33.$$

The elastic limit load, L, becomes from Eq. 37,

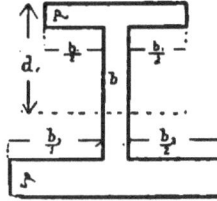

$$L = \frac{0.325 \,(8.05)^3 + 2.175 \,(1)^3 \,(3 \times 8.05 - 2 \times 1)}{3 \times 8.05 \times 132} \times 4 \times 36320 = 9924 \text{ lbs.},$$

and from Eq. 36,

$$L = \frac{4 \times 4 \,(3 \times 8.38 + 8.05) \, 30000}{3 \times 132} = 9924 \text{ pounds.}$$

The *elastic limit load* was between the applied loads 9493 and 11253 pounds from Mr. William Fairbairn's experiments, and their deflections were $0''.46$ and $0''.60$; the deflection required was $0''.33$.*

EXAMPLE 30.—Required the centre elastic limit transverse load of a Rolled Wrought-Iron Double T Beam when

Depth of T beam.......$d = 9''.44$, $C = 37165$ lbs. required for eq'brium,
Breadth of web ....$b = 0''.35$, $T = 30000$ pounds mean value,
   "   " top   flanges $b_1 = 2''.4$, $s = 10$ feet,
Depth   "   "   " $d_1 = 1''.0$, $m = 4$,
   "   " bottom   " $d_2 = 0''.44$, $d_c = 9.06$ assumed,
Breadth "   "   " $b_2 = 3''.95$, $d_\tau = 0.38$ "

---

* Fairbairn, "On Cast and Wrought-Iron," p. 102.

With these values the load, L, becomes from Eq. 28,

$$L = \frac{4 \times 4.3 \times 0.38 \ (2 \times 9.44 + 9.06) \ 30000}{3 \times 12 \times 10} \doteq 15218 \text{ lbs.,}$$

and from Eq. 38,

$$L = \frac{0.35 \ (9.06)^3 + 2.4 \ (1)^2 \ (3 \times 9.06 - 2 \times 1)}{3 \times 9.06 \times 12 \times 10} \times 4 \times 37165 = 15218 \text{ lbs.}$$

The elastic limit load was between the applied loads 14693 and 16373 pounds from Mr. William Fairbairn's experiments, and the deflection was $0''.35$ and $0''.45$ with these loads respectively.*

In Examples 29 and 30 the tensile elastic limit has been assumed to be 30000 pounds, the mean value for wrought-iron, and from it the position of the neutral line has been computed from Eq. 30, and then the required value of C to produce equilibrium.

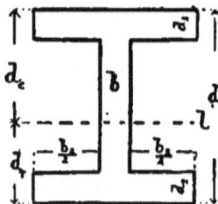

96. Double T *beams with tension and compression flanges equal in area.*

Example 31.—Required the centre transverse elastic limit load of a Rolled Wrought-Iron Eyebeam when

Depth of beam........$d = 9''.0$, $C = 63170$ lbs. required for eq'brium,
Breadth of web.......$b = 0''.6$, $T = 30000$ " mean value,
  "   " top flanges $b_1 = 4''.775, m = 4$,
Depth "   "   " $d_1 = 1''.0$, $s = 14$ feet,
  "   " bottom " $d_2 = 1''.0$, $d_e = 8''.2$,
Breadth "   "   " $b_2 = 4''.775$, $d_r = 0''.8$.

With these values the load, L, becomes from Eq. 37,

$$L = \frac{0.6 \ (8.2)^3 + 4.775 \ (1)^2 \ (3 \times 8.2 - 2 \times 1)}{3 \times 8.2 \times 12 \times 14} \times 4 \times 63170 = 26823 \text{ lbs.,}$$

---

* Fairbairn, " On Cast and Wrought-Iron," p. 103.

and from Eq. 28, in which $b = b + b_2 = 5.375$,

$$L = \frac{5.375 \times 0.8 \ (2 \times 9 + 8.2)}{3 \times 12 \times 14} \ 4 \times 30000 = 26823 \text{ lbs.}$$

The value of T is taken at the mean, from which d_x must be 0″.8 and $C = 63170$, that equilibrium shall exist between the moments of the tensile and compressive resistances.

The rolled eyebeam, of which the data given in Example 31 is the equivalent right line section, was tested by the Phœnix Iron Co. of Pennsylvania,* and its centre elastic limit load was found to be between the applied loads 26880 pounds and 28000 pounds; the deflection was 0″.572 and 0″.600 with these loads respectively.

The proportion of the load and area of the section that is sustained by each member is given in the following table:

	Load.	Area.
Compression Flanges	12.3	31.9 per cent.
" Web	37.7	32.9 "
Tension "	0.55	3.2 "
" Flanges	49.45	24.1 "
" " (practically lost)	—	7.9 "
	100.00	100.0

The *transverse elastic limit* load of the rectangular wrought-iron beam 5″.375 × 9″.0, from which the above described *eyebeam* may be supposed to have been cut, is from Eq. 27, when $C = 30000$ and $T = 30000$ pounds, the mean values for bar-iron,

$$L = \frac{4 \ (7.02)^2 \ 30000}{3 \times 12 \times 14} = 63076 \text{ pounds.}$$

From which it will be observed that the eyebeam, while containing only 31 per cent of the area of the rectangular beam,

---

* Phœnix Iron Company's "Handbook of Useful Information."

is able to sustain 42.5 per cent of its load, which is supposed to be due to the *elevation* of the *elastic* limit during the process of rolling, or the top of the beam must have been laid with steel.

SECTION IV.—*Circular Wrought-Iron and Steel Beams.*

97. Neutral Line. The position of the neutral line in Circular Wrought-Iron and Steel Beams for the different ratios of $C \div T = q$ required has been computed, also the factors f_c and f_x for use in Eqs. 76 and 77, and tabulated below for reference.

Ratio of Crushing to Tenacity, or $C \div T = q$.	Depth of Neutral Line Below the Crushed Side of the Beam, or d_c.	FACTORS FOR COMPUTING THE MOMENT OF RESISTANCE, RADIUS = 1.	
		f_c for Crushing Strain C.	τ for Tensile Strain T.
3.0	0.5375 d	0.6468	1.9399
2.875	0.5438 d	0.6642	1.9093
2.75	0.5505 d	0.6827	1.8766
2.625	0.5574 d	0.7024	1.8434
2.5	0.5646 d	0.7230	1.8084
2.375	0.5724 d	0.7458	1.7718
2.25	0.5804 d	0.7696	1.7314
2.125	0.5890 d	0.7952	1.6896
2.0	0.5980 d	0.8228	1.6454
1.875	0.6076 d	0.8526	1.5984
1.75	0.6178 d	0.8848	1.5482
1.625	0.6277 d	0.9198	1.4944
1.5	0.6404 d	0.9580	1.4368
1.375	0.6525 d	1.0000	1.3748
1.25	0.6666 d	1.0462	1.3074
1.125	0.6816 d	1.0976	1.2328
1.0	0.6976 d	1.1548	1.1548

From this Table the position of the neutral line in any wrought-iron or steel beam may be obtained by multiplying the diameter d, expressed in inches, by the decimal factor corresponding to the ratio $C \div T = q$. For ratios intermediate in value the position may be obtained by proportion.

98. Transverse Strength. This may be computed

from either Eq. 74 or 75, but with much less labor from Eq. 76 or 77, using the values of f_c and f_s given in the above Table.

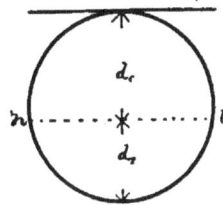

EXAMPLE 32.—Required the centre breaking or elastic limit load of a Circular Steel Beam when

The diameter $d = 1''.129$, $C = 41160$ lbs. mean of four tests,
" span.....$s = 20''.0$, $T = 39200$ " " " ten "
" factor...$m = \frac{4}{3}$, $q = 1.05$.

The position of the neutral line and factor, f_c, becomes, by proportion from the Table,

$$d_c = 0.6912 \times 1''.129 = 0.780, f_c = 1.132.$$

With these values the load, L, required becomes from Eq. 76,

$$L = \frac{4 (0.5645)^2 \, 1.132 \times 41160}{20} = 1676 \text{ pounds.}$$

From the United States Government Report of the Tests of Iron and Steel for the year 1885,* the elastic limit load of two of these beams that were tested was 1638 pounds with the values of C and T as given in the Example.

EXAMPLE 33.—Required the centre elastic limit load of a cylindrical Phœnix pin supported at both ends when

---

* Senate Ex. Doc. No. 36–49th Congress, 1st Session, p. 690.

The diameter.... $d =$ 2".5, $C =$ 30000 pounds from tests,
" span .... $s =$ 24".0, $T =$ 30000 " ' " "
" · factor...... $m = 4$, $f_c = 1.1548$ from the Table.

Then from Eq. 76 we have for the required load, ·

$$L = \frac{4 \, (1.25)^3 \, 1.1548 \times 30000}{24} = 11377 \text{ pounds.}$$

The elastic limit strength of this pin was 11000 pounds.
(Watertown Arsenal Report of Tests of Iron and Steel for
1881.)* The computed ultimate strength is 18795 pounds,
with C and $T = 50000$ pounds, the recorded ultimate strength
is 20000 pounds.

In order that these wrought-iron pins should *truly cross-break*, their deflection should not be less than 0.3024 d; in the
above examples the ultimate deflection was 1".278; while that
required for true cross-breaking was 0".75, the recorded deflection with 18000 pounds was 0".78.

A number of these pins, made of Phœnix and Pencoyd
iron, were tested with diameters ranging from 2½ to 5 inches,
but the span was so short, in nearly all cases, that the tensile
fibre strain was not fully developed, causing the transverse
elastic limit load to be not well defined and the ultimate load
to be larger than it should be. Though the computed elastic limit
loads do not differ very greatly from those determined by
tests, there is a very great difference between the computed
and experimental ultimate loads, the latter being the greater.

This series of tests illustrates the correctness of the statement made in Art. 24, that the *bending moment* of the applied load at the *inception* of the deflection of a beam is held
in equilibrium by the moment of a *purely* compressive resistance that is distributed over the section in an uniformly vary-

---

* House of Representatives Ex. Doc. No. 12, 1st Session, 47th Congress,
p. 171.

ing strain.   Test No. 741, page 186, of the Report, was of Phœnix iron 4″ in diameter and 24″ span.   The bending moment of the centre applied load is 6 L from Eq. 3, page 5, and the moment of compressive resistance is the product of the resultant (the area of the section by one half, C, the elastic limit compressive strength), by its lever-arm, $\frac{3}{8}d$, the distance of the centre of gravity of the pressure wedge below the axis; hence

$$6L = \frac{3d}{8} \times \pi \frac{d^2}{4} \times \frac{C}{2},$$

$$\therefore L = \frac{3\pi d^3 C}{384} = \frac{3 \times 3.1416 \, (4)^3 \, 30000}{384} = 47124 \text{ pounds.}$$

The observed elastic limit load was 48000 pounds with a deflection of 0″.0585, but as no correction appears to have been made for the settling of the beam on its bearings, we may conclude that the beam was on the verge of true deflection with its transverse elastic limit load.

SECTION V.—*Hollow Circular Wrought-Iron and Steel Beams.*

99. Neutral Line. The position of the neutral line in Hollow Circular Wrought-Iron and Steel Beams for the different ratios of $C \div T = q$ required has been computed, also the factors f_c and f_τ for use in Eqs. 91 and 92, and tabulated below.

Ratio of Crushing to Tenacity, or $C \div T = q$.	Depth of Neutral Line Below the Crushed Side of the Beam, or d_c.	FACTORS FOR COMPUTING THE MOMENT OF RESISTANCE, RADIUS = 1.	
		f_c for Crushing Strain C.	f_τ for Tensile Strain T.
3.0	0.7500 d	1.7320	5.1961
2.875	0.7597 d	1.7737	5.0992
2.75	0.7697 d	1.8172	5.0000
2.625	0.7800 d	1.8628	4.8898
2.5	0.7906 d	1.9105	4.7760
2.375	0.8014 d	1.9601	4.6566
2.25	0.8122 d	2.0108	4.5334
2.125	0.8241 d	2.0645	4.3928
2.0	0.8358 d	2.1251	4.2490
1.875	0.8478 d	2.1853	4.0966
1.75	0.8600 d	2.2479	3.9338
1.625	0.8727 d	2.3135	3.7586
1.5	0.8851 d	2.3823	3.5722
1.375	0.8979 d	2.4534	3.3728
1.25	0.9107 d	2.5267	3.1588
1.125	0.9233 d	2.6024	2.9296
1.0	0.9360 d	2.6806	2.6806

100. Transverse Strength. This may be computed from either Eq. 89, 90, 91 or 92.

EXAMPLE 34.—Required the centre breaking transverse load of a Hollow Wrought-Iron Cylindrical Beam, supported at both ends, when

The outer diameter... $d = 4''.0$, C and $T = 45000$ pounds,
" thickness of metal $t = 0''.1875$, $f_c = 2.6806$ from Table,
" span .... $s = 6'.0$ $m = 4$.

Then from Eq. 91 we have for the load,

$$L = \frac{4 \, (2)^2 \; 2.6806 \times 0.1875 \times 45000}{12 \times 6} = 5076 \text{ pounds.}$$

This " tube failed suddenly with 5824 pounds." (From " The Britannia and Conway Tubular Bridge," p. 435.)

CHAPTER VI.

TIMBER BEAMS.

101. The Compressive and Tensile Strength.
The great number of the different varieties of timber, and the great variation in the strength of the timber that is known in different parts of the globe by the same name, render the determination of *constants* or mean values for C and T for use in the computation of the strength of beams a very difficult matter, and we find that much difference exists between the *constants* that are given for the same timber by the older and more recent experimenters; especially is this the case for the crushing strength. The former make $C \div T$ less than 0.75 for all of the principal varieties, while the latter make it greater than *unity*.

The *computed transverse strength* of the beams broken by the older experimenters from their *constants* gives, practically, accurate results, or such as are within the limits of the variation in strength of the material, while those from the more recent experimenters make their *computed transverse strength* from twenty-five to one hundred per cent greater than their experiments gave, which indicates that with improved *testing* machines the more recent experimenters, such as Professor Thurston, Laslet and Hatfield, obtained their *crushing* strength at a point nearer the total destruction of the wood than the older with machines having less power.

102. To Compute the Compressive and Tensile Strength. *Crushing.*—From the fibrous character of timber and its weak lateral adhesion it is difficult to determine its crushing strength from direct pressure on small specimens, or that intensity of crushing strain that holds its tensile strength in equilibrium : this can be accurately computed from Eqs. 31 and 79, when the breaking, transverse and tensile strength have been determined from experiments.

EXAMPLE 35.—Required the Crushing Strength of Teakwood from the known tensile strength and the centre transverse breaking load of a rectangular beam when

Depth...$d=2.0$ ins., $T=15000$ lbs. mean of Mr. Barlow's tests,
Breadth .$b=2.0$ " $L=938$ "   "   "   "   "
Span....$s=7.0$ feet, $m=4$.

The position of the neutral line becomes from Eq. 30,

$$d_s = \frac{3 \times 2}{2} - \sqrt{\frac{9 \times 4 \times 2\,(2)^2\,15000 - 12 \times 938 \times 12 \times 7}{4 \times 4 \times 2 \times 15000}} = 0.35 \text{ ins.},$$

$$\therefore d_c = 2.0 - 0.35 = 1.65 \text{ inches,}$$

and the crushing value of C from Eq. 31,

$$C = \frac{3 \times 938 \times 12 \times 7}{4 \times 2\,(1.65)^2} = 10850 \text{ pounds.}$$

From Mr. Hodgkinson's experiments $C = 12100$ pounds for Teak.

Tensile Strength.—This may be computed from Eqs. 32 and 33, and from Eqs. 80 and 81, when the transverse and crushing strength have been ascertained from experiments.

RECTANGULAR WOODEN BEAMS.

103. Neutral Line. Table of positions of the neutral line in rectangular sections of wood, and the mean ultimate crushing and tensile strength of the different varieties of timber.

	Breaking Strength per Square Inch in Pounds.		Ratio of Crushing to Tenacity, $\frac{C}{T} = q.$	Depth of Neutral Line Below the Crushed Side of the Beam, or d_c.	Authorities.
	Crushing, C.	Tensile, T.			
Ash, American.....	4400	11000	0.4	0.8900 d	B., Barlow.
" "	5800	14000 `	0.4142	0.8877 d	H., Hodgkinson.
" English.......	8600 II	12000 M	0.7166	0.8267 d	M., Molesworth.
Beech, American ...	5800	15000	0.3866	0.8933 d	R., Rankine.
" " ...	6900	18000	0.3833	0.8941 d	
" English......	7700 II	9000 R	0.4142	0.8878 d	
" "	9300 II	12000 R	0.7750	0.8165 d	
Birch, American Black...........	7000	11600	0.6034	0.8477 d	
Birch, English.....	4530 II	11700 R	0.3872	0.8939 d	
" "	6400 II	15000 M	0.4266	0.8843 d	
Cedar, American Red	6000	10300	0.5825	0.8518 d	
Elm........... ...	6830	14000 M	0.4878	0.8706 d	
Fir, White Spruce..	6500 M	10000 M	0.6500	0.8376 d	
" " Christiana Deal....	5850 M	12000 M	0.4875	0.8711 d	
Oak, American Red	6000	10000	0.6	0.8495 d	
" " White	7200	18000	0.4	0 8900 d	
" "	9100	18000	0.5055	0.8673 d	
" English......	6500 II	10000 M	0.65	0.8376 d	
" "	9500 II	19000 M	0.5	0.8685 d	
Pine, Am. South- ern Long-leaf...{	8000	12600	0.6349	0.8410 d	
	8000	19200	0.4166	0.8525 d	
Pine, Am. White...	5000	10000	0.5	0.8685 d	
" Am. Yellow..	....	....	....	....	
" Red, European	5400 M	12000 M	0.45	0.8792 d	
" "	7500 M	14000 M	0.5357	0.8559 d	
" " Dantzic...	5400 M	8000 M	0.575	0.8533 d	
" " Riga	5748 H	11549 B	0.4977	0.8690 d	
" " "	6586 H	12857 B	0.5122	0.8660 d	
Poplar...........	5100 T	7000 T	0.73	0.8243 d	
Walnut, Black.....	....	....	...	....	

The position of the neutral lines in the above Table was computed from Eq. 25, d being *unity*.

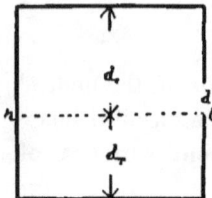

104. Transverse Strength. Timber beams breaking with a well-defined fracture in its fibres, the transverse strength may therefore be computed from the crushing and tensile strength by means of either Eq. 27 or 28, as the rectangular is the form that is principally used in wooden beams.

EXAMPLE 36.—Required the uniformly distributed breaking load of an American White Pine Rectangular Beam, when

The depth.....d = 14″.0, C =   5000 pounds mean of tests,
" breadth....b =   6″.0, T = 10000   "   "   "
" span......s = 28′.0, m = 8, Art. 34, and q = 0.5.

The position of the neutral line from the Table is

$$d_c = 0.8685 \times 14 = 12.16 \text{ inches,}$$

and the breaking load from Eq. 27 becomes

$$L = \frac{8 \times 6 \, (12.16)^2 \, 5000}{3 \times 12 \times 28} = 35200 \text{ pounds.}$$

The Table in Trautwine's Engineers' Pocket-Book gives 37800 pounds as the breaking load of this beam.

EXAMPLE 37.—Required the breaking load of a Rectangular English Oak Beam fixed at one end and loaded at the other, when

The depth.......d = 2″.0, C =   6500 pounds mean by test,
" breadth.....b = 2″.0, T = 10000   "   "   "
" span........s = 4′.0, q = 0.65 and m = 1, Art. 34.

The position of the neutral line is, from the Table,

$$d_c = 0.8376 \times 2 = 1.6752 \text{ inches,}$$

the load, L, from Eq. 27, becomes

$$L = \frac{1 \times 2 \, (1.6752)^2 \, 6500}{3 \times 12 \times 4} = 253 \text{ pounds.}$$

From Colonel Beaufoy's experiments the mean breaking load of 6 of these beams was 258 pounds each.*

EXAMPLE 38.—Required the centre breaking load of the beam described in Example 37, when

$$\text{The span} = 7.0 \text{ feet}, \ m = 4.$$

The required load, L, from Eq. 27, becomes

$$L = \frac{4 \times 2 \, (1.6752)^2 \, 6500}{3 \times 12 \times 7} = 578 \text{ pounds.}$$

From Mr. Barlow's experiments the mean breaking load of three of these beams was 637 pounds each.

EXAMPLE 39.—Required the centre breaking loads of the following Rectangular Wooden Beams, when

The depth ....$d = 2''.0$, $C =$ values from the Table.
" breadth.......$b = 2''.0$, $T = $ " " " "
" span.... .....$s = 50''.0$, $m = 4$.

The position of the neutral line is obtained by multiplying the depth, d, by the factor corresponding to the ratio $C \div T$ given in the Table, for each case.

Computed.   2 tests.

Christiana Deal..$L = \dfrac{4 \times 2 \, (1.7622)^2 \, 5850}{3 \times 50} = 969$.......940 and 1052 lbs.

English Ash.....$L = \dfrac{4 \times 2 \, (1.6534)^2 \, 8600}{3 \times 50} = 1254$.. ....1304 and 1304 "

English Birch. ..$L = \dfrac{4 \times 2 \, (1.7686)^2 \, 6400}{3 \times 50} = 1067$.......1164 and 1304 "

Am. Black Birch $L = \dfrac{4 \times 2 \, (1.6954)^2 \, 7000}{3 \times 50} = 1073$.......1027 and 1433 "

These experimental breaking loads are taken from Mr. P.

---

* Barlow's "Strength of Materials," p. 58.

W. Barlow's experiments;* the values for C and T are the mean that are usually quoted by authors.

EXAMPLE 40.—Required the centre breaking load of a Rectangular French Oak Beam, when

The depth...........$d =$ 7.5 inches, $C = 6000$ pounds,
" breadth..........$b =$ 7.5 " $T = 10000$ "
" span.............$s = 15.0$ feet, $q = 0.6$ and $m = 4$.

The position of the neutral line from the Table becomes

$$d_c = 0.8495 \times 7.5 = 6.374 \text{ inches,}$$

and the required load, L, from Eq. 27, becomes

$$L = \frac{4 \times 7.5 \, (6.374)^2}{3 \times 12 \times 15} \, 6000 = 13543 \text{ pounds.}$$

M. Buffon, in experiments made for the French Government,† broke two of these beams with 13828 and 14634 pounds respectively; from Rondelet's experiments the values of C and T are supposed to be equal to those given in the Table for American Red Oak.

CIRCULAR WOODEN BEAMS.

105. Neutral Line. The position of the neutral line and the value of the factors f_c and f_τ for use in Eqs. 76 and 77 may be obtained from the following Table, by proportion when necessary:

---

* Barlow's " Strength of Materials," p. 86. † *Ibid.*, p. 56.

Ratio of Crushing to Tenacity, or $C \div T = q$.	Depth of Neutral Line Below the Crushed Side of the Beam, or d_c.	FACTORS FOR COMPUTING THE MOMENT OF RESISTANCE, RADIUS = 1.	
		f_c for Crushing Strain, C.	f_τ for Tensile Strain, T.
1.0	0.6976 d	1.1548	1.1548
0.9829	0.7000 d	1.1632	1.1433
0.9473	0.7050 d	1.1812	1.1189
0.9126	0.7100 d	1.1991	1.0943
0.8790	0.7150 d	1.2174	1.0708
0.8463	0.7200 d	1.2357	1.0458
0.8133	0.7250 d	1.2540	1.0200
0.7837	0.7300 d	1.2726	0.9974
0.7541	0.7350 d	1.2912	0.9732
0.7245	0.7400 d	1.3098	0.9490
0.6965	0.7450 d	1.3287	0.9251
0.6686	0.7500 d	1.3476	0.9013
0.6423	0.7550 d	1.3666	0.8775
0.6161	0.7600 d	1.3856	0.8538
0.5912	0.7650 d	1.4048	0.8304
0.5664	0.7700 d	1.4240	0.8066
0.5429	0.7750 d	1.4434	0.7832
0.5194	0.7800 d	1.4628	0.7598
0.4971	0.7850 d	1.4828	0.7365
0.4749	0.7900 d	1.5020	0.7133
0.4534	0.7950 d	1.5216	0.6895
0.4320	0.8000 d	1.5412	0.6658
0.4129	0.8050 d	1.5611	0.6442
0.3938	0.8100 d	1.5810	0.6226

106. Transverse Strength.

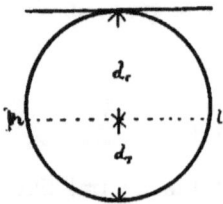

The transverse strength of circular wooden beams may be computed from either Eq. 74 or 75, also from Eq. 76 or 77, with the aid of the values of the factors f_c and f_τ deduced from the above Table.

EXAMPLE 41.—Required the centre breaking load of a Circular Christiana Deal Beam, when

The diameter $d = 2''.0$, $C = 5850$ pounds mean of tests,
" span ... $s = 48''.0$, $T = 12000$ " " " "
$\qquad m = 4,$ $\qquad q = 0.4875.$

The required load, L, becomes from Eq. 77,

$$L = \frac{4 \ (1)^2 \ 0.726 \times 12000}{48} = 726 \text{ pounds.}$$

Mr. Barlow broke three of these beams with 740, 796 and 780 pounds respectively, the mean being 772 pounds.*

107. Relative Strength of Square and Circular Timber Beams. The required relation will be obtained from either Eq. 83, 84 or 85, as the case may require.

EXAMPLE 42.—Required the centre breaking load of a Circular Christiana Deal Beam from that of the circumscribed square beam of the same span and material, when

Side of the square...$d = \ $ 2".0, $C = \ $ 5850 lbs. mean of tests,
Diameter of the circle $d = \ $ 2".0, $T = 12000$ " " " "
Span ....$s = 48$".0, $q = 0.4875$,
f, from the Table, page 108, becomes 0.8711 when $d = 1$,
and f_c, from the Table, page 112, becomes 1.4912 when $r = 1$.

With these values Eq. 83 becomes

$$\frac{3 f_c}{8 f^3} = \frac{3 \times 1.4912}{8 \times (0.8711)^3} = 0.737;$$

hence, $Circle = Square \times 0.737.$

Breaking strength of 2" sq. beam, Mr. Barlow's tests, 1117 lbs.
"   "   " 2" circ. " 1117×0.737 ...... 823 "
Mean breaking strength of 3 beams, Example 39.... 772 "

Mr. Barlow says that the 2" square and the 2" diameter circular beams were cut from the same plank, "which was a very fine specimen of Christiana deal." The breaking strength of the 2" square and 48" span beam, with the above values of T and C, should have been 986 pounds, and the circular beam $986 \times 0.737 = 726$ pounds, as in Example 41.

* Barlow's "Strength of Materials," p. 78.

CHAPTER VII

108. General Conditions of Failure of Columns. Euler and Tredgold are credited with the only well-recognized attempts that have been made to deduce rational formulas for the strength of columns, but the basis of each of their theories involves the assumption of the existence of conditions that render their application to the actual phenomena observed in practice inapplicable without the aid of empirical factors determined from experiments.

The laws that govern the breaking strength of pillars were investigated, experimentally, by Mr. Eaton Hodgkinson in 1840, but no *rational* theory has ever been advanced that will explain the phenomena observed by him and subsequent investigators; and our *constructors* are to this time using the empirical rules of either Hodgkinson or Gordon, that were deduced from the former's experiments, to compute the requisite dimensions of their pillars, or as they may have been modified to conform to the results obtained by subsequent investigators from new material or that produced from improved methods of manufacture.

The effect that will be produced upon a given piece of material, when subjected to a strain in the direction of its length, depends entirely upon its deflection. The deflection varies with the material and with the ratio of the length to the least diameter. It has been found, experimentally, according to Mr. Tredgold, " that when a piece of timber is compressed in the direction of its length, it yields to the force in a different manner according to the proportion between its length and the area of its cross-section," which is according to the amount it is able to bend or deflect laterally.

The material of a pillar when subjected to an applied load in the direction of its length, in order to avoid the strain thus brought upon its fibres, deflects laterally and assumes a form composed of one or more curves, as its ends may be round or flat ; a round end pillar assumes the *one* curve form and the flat end pillar takes a form made up of *three* or *four* curves. The former is represented in Fig. 34, and will be called the Triple Flexure form ; it is that in which the pillar offers the least resistance to breaking by an applied load.

When a pillar of the Triple Flexure form fails with deflection, tension exists in its fibres at g (Fig. 34), and compression at m ; hence there must be a point in the line gam, at which there is no strain ; likewise there must be a point without strain in the line gds ; and compression existing in the fibres at the points o, f and k, the entire side, $kefbo$, must be in like condition. The fibre strains at two sections of the pillar, such as ac and db, must therefore increase uniformly in intensity from zero at a and d, to its greatest at c and b, respectively. The failure of a few columns along the lines ac and db indicates that they join the points of reverse curvature of each side of the pillar, though a fracture beginning at c may follow the line of less resistance, cm. There is also a point, n, in the line g f, at which the strain is zero in intensity, and a curved line, a n d, connecting these zero points is the *neutral surface* of the column.

Since the column must bend symmetrically there is a point in each of the lines ac and db that is one fourth the length of the column from each of its ends, mk and so, respectively, thus making the middle curve of the *triple flexure* form *one half* of the length of the column. The angle that the lines ac and db make with the plane of the ends varies with the material, but in all pillars it is supposed to approximate the angle of 45 degrees.

However short a pillar may be when loaded, it will attempt to assume the *triple flexure* form. The curves above and below the lines *ac* and *db* being the first to take their shape, thus give rise to the various phenomena that are observed when *short* blocks of granular textured material, such as cast-iron, are crushed. When the points *c*, *f* and *b* coincide, the block is sheared across at one plane; when *a* coincides with *m* and *d* with *s* the block splits up into four or more wedges.

The strength of a pillar generally diminishes as it becomes longer in proportion to its least diameter. Should a series of pillars of an uniform section be constructed with progressively increasing lengths and the strength of each be obtained by experiment, there would be found one length whose strength per square inch of section would be greatest; the strain must be uniformly distributed over the section of the pillar and is the *crushing* value of *C* for the material. As the length increases the strength decreases, until a length of pillar is reached whose strength is just one half of the greatest strength above described; the strain must be *uniformly varying* in intensity—zero at *g* and greatest at *f*, where its intensity per square inch is the crushing value of the material. The pillar of our series, with this ratio of length to its least diameter, will fail by *crushing* without deflection when it is on the *verge* of failing by crushing with deflection; *twice* the mean intensity of its strength per square inch of section is the crushing strength of the material, or that *crushing strength* that equilibrates the *tenacity* when beams and columns fail by *cross-breaking*.

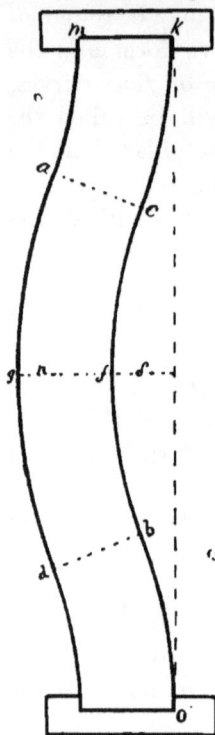

Fig. 34

At the *inception* of the deflection of the pillars of our series that fail *with* deflection, the *neutral line* coincides with the side, *mgs*, of the pillar ; as its length continues to increase it will now be able to deflect, all compression having been removed from one side, and the neutral line will move from *g* toward *f*, a distance equal to the deflection, until the deflection, *gn*, equals that depth of *extended* area that was found to be necessary for equilibrium when a beam of the same material and section is broken by a *transverse* load, and the pillar will fail with all of the phenomena of true *cross-breaking* in a beam.

From the length at which a pillar of our series just begins to deflect, to that at which it deflects sufficiently to *cross-break*, the anomaly of the strength of pillars, increasing with an increase of length, is exhibited ; of two pillars of the same section and material, the one that deflects most, if it breaks within the limits of deflection above described, will sustain the greater load. When the length is increased beyond that at which the pillar first breaks by true cross-breaking, its strength again diminishes, and the further anomaly is presented of a column failing by cross-breaking with the *same* load that a column of the same section and shorter length will fail with by crushing, but with less deflection.

The pillar by deflecting seeks to relieve itself of the *resting* of the load *directly* on its middle section, but this it can only do to the extent of $gn = d_\tau$, or that value required for rupture in beams, and it must fail by the direct pressure of the load along the line *nf*, as it is being transmitted to the foundation, *so*, for all lengths of pillars ; but when it deflects so much that the resultant of the applied load passes without the middle section of the pillar, it thus increases its effect by a *cantilever* strain, which effect must be deducted from the amount of the *two pressure wedges* that are required along *nf* to rupture the fibres by both tension and compression, as explained in Art. 109.

Pin End Columns.

The "failure" of long pin end columns or those that fail by *cross-breaking*, presents some interesting peculiarities. The *frictional resistance* offered to the movement of the pillar around the pin, as it deflects, increases with the size of the pin; it varies in pillars of like dimensions and material when tested with pins of the same size, and decreases, in all cases, after the pillar has been fairly set in motion by the lateral deflection. Thus, the results of experiments show that a load less than that required to produce fracture by compression will be *sustained*, but as the deflection approaches near to d_r in value, the amount of friction becomes so small that the pillar will "suddenly spring" to a deflection that will cause the pillar to fail with this load by true cross-breaking, and of two like columns, one will *sustain* for a time a greater load than the other, on account of greater *frictional resistance* around the soffit of the pin, evidenced by its deflection being the least; the greater load will then commence to decrease, with a freer motion of the pillar, and will finally reach a point where it will "suddenly spring" to a *cross-breaking* deflection, thus breaking with a less load than it had already sustained, but with less deflection. That the pillar is truly broken by its "sudden spring" is evidenced by the fact that upon being released from the pressure of the load it will now require with wrought-iron pillars, in many cases, less than one half of the original load to produce the greatest deflection before obtained. This uncertain and variable frictional resistance around the pins of pin end columns gives rise to many anomalous results that would not otherwise be encountered in making a series of tests with pin end pillars when made of homogeneous material, such as wrought-iron, and renders it impossible to determine the exact law that governs their failure without eliminating its irregular effect. The *momentum* of the *sudden spring* frequently causes the observed deflection to greatly exceed the true deflection at failure, and the load *sustained* at the instant of springing often exceeds the true cross-breaking load. These sources of erroneous observation become less in effect as the size of the pin is increased and as the pillar becomes shorter in length, and to increase in effect with an increase in the length of the pillar and a decrease in the size of the pin.

109. Resistance of the Cross-Section of the Column. From the preceding Art. it is apparent that when a pillar fails by crushing without deflection, its strength is simply the product of its area by the mean intensity of the pressure per square inch, which varies with its length, and that after it begins to deflect, equilibrium must be established between the *moments* of the *applied* and *resisting* forces with reference to the *fulcrum* or origin of moments, f (Fig. 34), the *neutral line* being at n, the line of direction of the forces, perpendicular to the section fg, and uniformly varying in intensity.

When a column fails with a deflection that compels the resultant of the applied load to pass through the middle section of the pillar, it will only be necessary to find the amount of *direct* pressure along the line nf that will cause the pillar to "fail." The *tensile* strain along the line gn must be held in equilibrium by an *uniformly varying* crushing strain along the line nf; the maximum intensity of this *pressure wedge* is always less than the crushing value of C for the material, until $gn = d_\tau$, which is required for rupture in a beam, when it becomes equal to it. It will also require another *pressure wedge* along this same line nf to *crush* the material, independently of the *pressure wedge* that is held in equilibrium by the *tensile* strain along the line gn; hence the actual pressure required along the line nf is the sum of the two pressure wedges, *one* that *crushes* the material direct, and the *other* that ruptures the *extended* portion of the pillar ng. When $gn = d_\tau$, these two *pressure wedges* become equal in size, and they must be constant in value for all lengths of pillars that fail by true *cross-breaking*, and their sum is the breaking load of the pillar as long as the resultant of the applied load passes through the middle section of the pillar.

The uniformly distributed load on the top of the pillar at *mk*, by the deflection, has been converted into an *uniformly varying* load at the section *fg*; it will so continue when the pillar becomes a cantilever, being zero in intensity next the concave side of the pillar and greatest at the tangent line connecting the points *k* and *o*, and its *moment* or power to break the pillar as a cantilever must be computed with reference to the fulcrum *f*, its lever-arm will be $(\delta - \frac{1}{3}d)$ for the rectangle, $(\delta - \frac{3}{8}d)$ for the solid circle, and $(\delta - \frac{1}{4}d)$ for hollow circles, in which *d* is the least diameter and δ the deflection. The direct tensile and compressive strain along the line *gf* that is required to hold in equilibrium the *moment* of the applied load acting upon the pillar as a *cantilever* must be deducted from the sum of the two pressure wedges that are required to cross-break the pillar, in order to obtain its true breaking load.

Columns can be constructed of wrought-iron, steel, wood, and some cast-iron that will "fail" by *cross-breaking*, with the full value of the sum of these two equal pressure wedges; but with most cast-iron and other material in which $C \div T$ is approximately greater than 1.75 for solid circles, 3.00 for hollow circles and 2.0 for rectangles, the pillar becomes a cantilever before it deflects sufficiently to cross-break.

The sum of these equal pressure wedges is the greatest load that a pillar can bear at the instant of "failure by cross-breaking," but experimenters frequently obtain for *short* pillars a greater *apparent* breaking load, from the fact that as the resultant of the load passes through its middle section, the pillar cannot at the *instant* of true *failure* escape from under it, and new conditions are set up, such as making two pillars, etc.

The formulas deduced for failure of pillars by *cross-break-*

ing from this theory are *identical* in form with those given by Mr. Lewis Gordon, and it can thus be seen why, that although his formulas are deduced from impossible conditions of cross-breaking, they can be so modified by the teachings of experiment that very accurate results may be obtained from their use.

110. Notation. The following notation has been adopted from previous chapters, and will have the same significance wherever it is used. Other special notation will be given as it may be required.

C = the greatest intensity of the compressive strain in lbs. per square inch.
T = the   "   "   " tensile   "   "   "   "   "
q = the quotient arising from dividing C by T.
b = the width of the section in inches.
d = the depth in the direction in which flexure takes place in inches.
d_e = the depth of the compressed area in inches.
d_r = the   "   " tension   "   "   "
L = the breaking load in pounds.
l = the length of the pillar in inches.
δ = the deflection in inches.

111. Deflection of Columns. In the present state of our knowledge of the cause of *flexure*, the laws that determine the amount that a pillar will deflect at the instant of failure can only be obtained from a carefully conducted series of experiments. On account of the expense and the great pressure required, the number of experiments on the breaking strength of pillars of the size used in structures is very small and incomplete for all lengths required for a complete knowledge of the subject; and as the theorists have taught, following Euler, that the strength of a column is independent of its deflection, it has caused the experimenters to so regard it and pay it but little attention, their means of measuring it being crude, and the record made in an unsatisfactory manner.

The amount of deflection that is required, before a column

can be broken, is, comparatively, a very small quantity, much less than that required for rupture should the pillar be broken as a beam, and requires very accurate means for its measurement. The amount that a pillar deflects is usually stated to be the distance that its middle section "moves," measured from a given stationary point, which is only correct when the points k, f and o (Fig. 34) lie in the same right line at the beginning of the movement; should f lie to the right of the line ko, the observed deflection will be too great by its distance from the line, and when f lies to the left of the same line ko, the observed deflection will be too small by its distance from the line ko, which must be parallel to the line of direction of the resultant of the applied load. The deflection is an incidental quantity ; the distance required is the *lever-arm* of the resultant of the load, which is the perpendicular distance from the fulcrum, f, to its line of direction.

l = the length of the pillar in feet and decimals,
d_c = the depth of the compressed area in inches,
c = a constant quantity determined from experiments,
δ = the breaking deflection in inches.

Then

$$\delta = c\,\frac{l^2}{d_c}. \tag{93}$$

The factor, c, must be determined for the different material used in structures, from experiments on fixed, round, two pin, and on one pin and one fixed end columns ; it has a different value for each material and style of end connections, and probably for each different shape of cross-section.

EXPERIMENTS.

The experiments required to determine the laws that govern the deflection of columns of a given section and material may be very much simplified by determining definitely the following points :

1st. The ratio of $l \div d$ that gives the true crushing value of the material.

2d. The *largest* ratio of $l \div d$ that sustains its greatest load without deflection.

3d. The *smallest* ratio of $l \div d$ that *cross-breaks*.

4th. The *largest* ratio of $l \div d$ that cross-breaks without *leverage* of the applied load.

The law governing the change in the strength from the 1st to the 2d and from the 2d to the 3d ratio would then be required; from the 3d to the 4th ratio the strength would be constant, and for larger ratios than the 4th the deflection, or lever-arm of the load, would be the only varying element, and only for these pillars will it be necessary to determine the values of the factor c, in Eq. 93, for the various kinds of material used in structures.

No attempt will be made to determine, for any given material, the limits above defined, from the incomplete record of experiments at present existing.

112. Classification of Pillars. From the description of the manner of "failure" of columns of various lengths, it is evident that they may be divided into the following general classes:

1st. PILLARS THAT FAIL BY CRUSHING.

2d. PILLARS THAT FAIL BY CROSS-BREAKING.

These classes may be again divided into Cases, which, for convenience of reference, will be numbered consecutively.

Pillars that fail by crushing.

Case I.—PILLARS THAT FAIL WITH THE FULL CRUSHING STRENGTH OF THE MATERIAL.

Case II.—PILLARS THAT FAIL WITH LESS THAN THE FULL CRUSHING STRENGTH OF THE MATERIAL AND WITHOUT DEFLECTION.

Case III.—Pillars that fail with less than the full crushing strength of the material and with deflection.

Pillars that fail by cross-breaking.

Case IV.—Pillars that cross-break from compression.

Case V.—Pillars that cross-break from compression and cantileverage.

End Connections.

Columns are again classed from the form or shape of their end connections, while they all fall under one or the other of the five cases given above. In columns of a given material and dimensions, the only element of variation in their strength arising from differences in its end connections is the *deflection*.

A *flat or fixed end column* is one whose ends are planes at right angles to its length; for a given material and dimensions it deflects less, and is, therefore, the form of *greatest strength*.

A *round end column* is one whose ends are spherical; consequently, being free to move laterally, it deflects most and is the form of *least strength*.

Pin end is a class of columns used in bridge construction in which the load is applied to pins passing through holes in its head and foot, and by them transmitted to the column. The smaller the pin and the less frictional resistance that may be developed between the *pin* and its *soffit*, the nearer it approaches the condition of a true *round* end column, and the less strength it will exhibit, while the larger the pin and the greater frictional resistance that may be developed the nearer it approaches the condition of a *flat* end pillar, and the greater strength it will exhibit.

One pin and one flat end, in its strength, follows laws similar to the above and a mean between flat and round end pillars.

MATERIAL AND SECTION.

In practice, columns are again classed as wooden, cast-iron, wrought-iron and steel, from the material of which they are made, and from the shape of their cross-sections into rectangular, circular, hollow circular, angle-iron, box, channel, eye-beam, tee, etc. From the incomplete tests at present made the lengths of columns of a given section and material that belong to each of the five cases of failure given above cannot be determined. In the examples quoted in the sequel these limits are, however, well defined in some special cases of material.

113. Case I.—COLUMNS THAT FAIL WITH THE FULL CRUSH-ING STRENGTH OF THE MATERIAL.

Let A represent the area of the section of the pillar in square inches, then as the load must be *uniformly distributed* for this case, the crushing load will be

$$L = A \times C. \qquad (94)$$

114. Case II.—COLUMNS THAT FAIL WITH LESS THAN THE FULL CRUSHING STRENGTH OF THE MATERIAL AND WITHOUT DE-FLECTION.

In *all* pillars of this class, at failure, the strain at the surface of one side is constant and equal to the *crushing* value of C for the material ; the strain at the surface of the opposite side varies with the length, from the crushing value of C to zero in intensity, at which length the strain is uniformly varying, the mean intensity being $C \div 2$.

$l_0 =$ the length of pillar that fails with full crushing strength,
$l =$ the   "   "   "   "   "   " a mean intensity, $C \div 2$,
$x =$ the   "   "   " whose strength is required,
$A =$ the area of the section of the pillar in square inches.

The values of l_0 and l vary with the material and are determined experimentally ; *assuming* that the intensity of the strain on one side of the pillar varies with the length, then for the length x it will be $C\left(\dfrac{l-x}{l-l_0}\right)$ and the mean for the section will be $\left(C + C\left(\dfrac{l-x}{l-l_0}\right)\right) \div 2$, and the load that the pillar of the length x will fail with is

$$L = A \times C\left(\frac{2l - l_0 - x}{2(l - l_0)}\right). \qquad (95)$$

For the *longest* column to which this formula applies, $l = x$, and Eq. 95 becomes

$$L = \frac{A \times C}{2} \qquad (95\text{A})$$

To Compute the Crushing Strength.—The above described manner of failure of pillars and the formulas resulting may be used very advantageously to compute the *crushing* strength of any material, which, in this case, is not affected by the tensile strength. And we are thus furnished with a valuable means of testing the correctness of the computed crushing strength for the same material when broken in a beam by the methods given in Articles 39 and 55; also that obtained from direct experiments with short blocks.

Deducing the value of C from Eq. 95A, we have

$$C = \frac{2 \times L}{A}, \qquad (95\text{B})$$

which is the formula required from which to compute the crushing strength of the material in a pillar that fails or sustains its greatest load *without* deflection when it is on the *verge* of failing *with* deflection.

EXAMPLE 43.—Required the greatest strength of a Rectangular Wrought-Iron Column, tested with two $1''\frac{1}{2}$ pin ends, when the deflection $\delta = 0$, area $A = 8.85$ square inches, and

$C = 50000$, assumed to be equal to the tensile strength obtained from direct tests.

From Eq. 95A we have

$$L = \frac{8.85 \times 50000}{2} = 221250 \text{ pounds.}$$

From the United States Government Watertown Arsenal Tests, 1882–1883,[*] the greatest strength of nine of these columns was found by experiments, the *mean* was 234850 pounds, the deflection varied from $0''.02$ to $0''.12$, the mean was $0''.07$; the lengths varied from 54 to 78 inches, and they were all approximately three inches square.   The mean strength of four of these columns, tested with one *flat* and one $1''\frac{1}{4}$ pin end was 210800 pounds, the mean deflection, $0''.12$; the length of two was 90 inches and that of the other two columns, 120 inches.

The mean strength of four of these columns, tested with *flat* ends, was 217875 pounds, the mean deflection was $0''.13$, the lengths were 90 and 120 inches.

115. Case III.—Columns that fail with less than the full crushing strength of the material and with deflection.

The pillars of this class vary in length for any material and section, between the limits of l of Case II., or that length at which the mean intensity of the crushing load is $C \div 2$, and that length of Case IV. that first fails by cross-breaking, both limits being determined by experiment.

$T' = $ the greatest tensile strain at the middle section.
$C' = $ the   "   compressive strain required to balance T'.

Then assuming that the intensity of the tensile strain in the convex side of the pillar varies with the deflection, δ, from 0

---

[*] Senate Ex. Doc. No. 5—48th Congress, 1st Session, pp. 60–67.

to T, the *tenacity* of the material, the following formulas are deduced for the various sections:

$$C' = qT', \quad T' = \frac{\delta T}{d_{\tau}},$$

$$\therefore C' = q \cdot \frac{\delta T}{d_{\tau}}.$$

By giving to q and δ the proper value for the section, the following formulas have been deduced:

RECTANGULAR PILLARS.

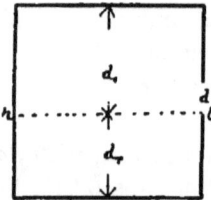

From Eqs. 27 and 28, page 38, we have

$$q = \frac{d_{\tau}(2d + d_{c})}{d_{c}^{2}}, \text{ in which } d_{\tau} = \delta.$$

Then the expression for C' becomes, from substituting these values,

$$C' = \frac{\delta^{2}(3d - \delta) T}{d_{\tau}(d - \delta)^{2}}.$$

The sum of the greatest intensities of the two pressure wedges at "failure" is $C + C'$.

$$\therefore L = \left[C + \frac{\delta^{2}(3d - \delta)}{d_{\tau}(d - \delta)^{2}} \cdot T \right] \frac{b(d - \delta)}{2}. \quad (96)$$

The value of d_{τ} in this formula is that required for rupture of the column when broken as a beam.

EXAMPLE 44.—Required the 'greatest strength of a Rectangular Wrought-Iron Column, tested with two $1''\frac{1}{2}$ pin ends, when

The diameter.......$d = 3''.00$, $C = 50000$ pounds assumed,
"   breadth........$b = 3''.00$, $T = 50000$ "   by test,
"   deflection ......$\delta = 0''.34$, $d_{\tau} = 0''.66$ for rupture.

From Eq. 96 we have

$$L = \left[50000 + \frac{(0.34)^2 (3 \times 3 - 0.34)}{0.66 (3 - 0.34)^2} \cdot 50000 \right] \frac{3 (3 - 0.34)}{2} = 242258 \text{ pounds.}$$

The greatest strength was 284000 pounds,[*] the length was 30 inches.

CIRCULAR PILLARS.

By substituting $C + C'$ for $2C$ in Eq. 100, page 134, which gives the resultant of the load for this case, we have for the load that the pillar will fail with,

$$L = \left[C + \frac{q \delta T}{d_{\tau}} \right] r^2 f. \tag{97}$$

The numerical values required for the factors q and f are to be taken from the Table, page 134, for that position of the neutral line that corresponds to $d_c = d - \delta$. The value of d_{τ} is that required for rupture of the material in a circular beam, with its full compressive and tensile strength.

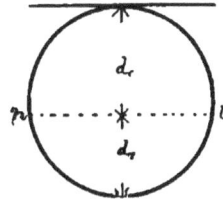

EXAMPLE 45.—Required the greatest load sustained by a Cylindrical Column of Midvale Steel, when

The diameter $d = 1''.129$, $C = 152000$ pounds mean of tests,
  " length....$l = 8''.96$, $T = 112285$ "   "   "   "
  " deflection $\delta = 0''.25$, $q = 1.353$.

From the Table, page 134, $f = 1.1839$, for the neutral line at failure, $d_c = 1''.129 - 0.25$, $q = 0.5278$, and $d_{\tau} = 0.38$, the position of the neutral line of *rupture*.

---

[*] From the United States Government Watertown Arsenal Tests, 1883, Senate Ex. Doc. No. 5—48th Congress, 1st Session, page 56.

From Eq. 97 we have

$$L = \left(152000 + \frac{0\;25 \times 0.5278 \times 112285}{0.38}\right)(0.5645)^2\;1.1839 = 72056 \text{ pounds.}$$

The greatest strength was 81250 pounds.[*]

HOLLOW CIRCULAR PILLARS.

By substituting $C + C'$ for $2C$ in Eq. 101, page 135, which gives the resultant of the load for this case, we have

$$L = \left[C + \frac{q\delta T}{d_{\tau}} \right] rtf_0,\qquad (98)$$

in which the numerical values required for q and f_0 are to be taken from the Table, page 136, for that position of the neutral line at failure that corresponds to $d_c = d - \delta$, the radius of the outer circle being r and t the thickness of the metal, both to be expressed in inches. The numerical value of d_τ is that required for rupture of the material in a hollow circular beam, with its full compressive and tensile strength.

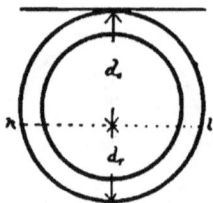

116. Case IV.—COLUMNS THAT CROSS-BREAK FROM COMPRESSION.

The shortest pillar of this class, for a given section and material, is that (Fig. 34) in which $gn = \delta = d_\tau$, or that depth of extended area required for equilibrium when a beam of this section and material is broken by a transverse load, and the longest is that that just deflects sufficiently to allow the resultant of the applied load, L, to pass through the *fulcrum, f*; this varies with the section, for a given material.

---

[*] From the United States Government Watertown Arsenal Tests for 1883–84, Senate Ex. Doc. No. 35—49th Congress, 1st Session, p. 369.

The load is the *sum* of the two pressure wedges. In rectangular areas or those that may be divided into rectangular areas, the load is the compressed area, multiplied by the *crushing* strength of the material.

Rectangular Pillars.

The position of the *neutral line* must be computed from either Eq. 25 or 26, page 38. Then, in order that the breaking load of a given pillar may be deduced from the principles applicable to Case IV., δ, the deflection, must be equal to or greater than d_e, and equal to or less than $\frac{1}{3}d$.

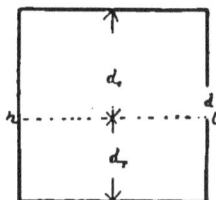

Then

$$L = bd_e C, \tag{99}$$

from which the required load may be computed, being the sum of the two pressure wedges required for rupture.

EXAMPLE 46.—Required the breaking load of a Rectangular Yellow Pine Column, tested with flat ends, when

The diameter..$d = 5''.5$, $C = 5230$ pounds mean of four tests,
" breadth..$b = 5''.5$, $T = 15478$ "   "   " tests,
" deflection $\delta = 0''.66$, $q = 0.337$.

The position of the neutral line, computed from Eq. 26, page 38, is $d_e = 5''.39$, and $d_r = 0.11$, being less than the deflection, the column failed by cross-breaking; the deflection being less than one third of the least diameter, it failed without cantileverage.

$$\therefore L = 5.5 \times 5.39 \times 5230 = 155043 \text{ pounds.}$$

In a series of experiments to test the strength of yellow pine columns with flat ends, conducted on the United States Government testing machine, at Watertown Arsenal, 1881–2,[*] thirty columns were broken, whose lengths varied from $l = 27d$ to $l = 45d$, which appears to be the limits of *cross-breaking*, without leverage for the yellow pine tested. The *crushing* strength obtained from short blocks differed so greatly that a *mean* was not admissible, though two consistent classes can be made of the material from these tests. The mean of four tests was $C = 5230$ pounds, from which the required load in the above example was computed ; the mean load from three tests was 154000 pounds. The mean value of C from three other tests was 3600 pounds, from which the computed load in the following Table was obtained. The tensile strength was obtained on the same machine the previous year, and not from the material composing the columns tested.

The " experimental " load in the Table is a mean of the number of tests given in the last column ; from " failure at knots and diagonally," the remainder of the thirty experiments mentioned above were rejected.

$l \div d$.	DIMENSIONS IN INCHES.			Deflection in Inches.	LOAD IN POUNDS.		No. of Tests.
	l.	b.	d.		Computed.	Experiment'l	
27.0	180	15.6	6.6	.5 to .62	344822	409000	3
30.8	180	12.0	5.8	.73 " 1.26	235000	250000	1
31.2	240	9.67	7.7	.82 " 1.6	249253	281000	2
31.2	180	15.5	5.6	.52	290718	344000	2
32.8	180	5.5	5.5	.52	101574	129500	2
36.0	180	12.1	5.0	1.3 " 1.8	202554	230000	1
38.2	210	5.5	5.4	1.03 " 1.95	100584	97330	3
40.0	180	11.6	4.4	.9 " 1.30	170800	126350	2
43.0	320	9.28	7.4	.99 " 2.55	231516	199830	3
45.0	240	5.4	5.4	1.28	98776	84900	2
45.0	180	.11.35	4.1	.95	166394	142000	1

In the above Table the second and sixth tests carried the maximum load with deflections that varied a half an inch in amount before failure.

---

[*] Senate Ex. Doc. No. 1—47th Congress, 2d Session, p. 321.

EXAMPLE 47.—Required the breaking loads of the series of White Pine Columns tested with flat ends, on the United States Government machine, at Watertown Arsenal, 1881–2,* when $T = 10000$ pounds mean of other tests, $C = 2500$ pounds mean of five of these tests. The position of the neutral line from Eq. 26 for $q = 0.25$ is

$$d_c = 0.92d.$$

The breaking loads in the following Table were computed from Eq. 99, as in Example 46; the first pillar of the Table did not deflect quite sufficiently to belong to this class of columns, but as the error is small, the lower limit is taken to be, as in yellow pine, $l = 27d$; the greater limit was not determined by the tests.

The "experimental loads" and the dimensions are the *means* of those given for three tests.

	DIMENSIONS IN INCHES.			Deflection in Inches.	LOAD IN POUNDS.	
$l \div d.$						
	$l.$	$b.$	$d.$		Computed.	Experimental.
27	180	15.6	6.66	.4 to .3	239400	227300
32	180	15.6	5.62	.37 " .83	202800	164100
32	240	9.4	7.45	.7 " 1.2	161200	170300
33	180	11.3	5.4	.6	165400	156200
36	280	9.6	7.69	.66 " 1.2	169900	153300
40	180	11.6	4.48	.43 " 1.25	119700	130700
43	320	9.3	7.47	.7 " 1.67	159700	147600

CIRCULAR PILLARS.

For this Case the deflection, δ, must be equal to or greater than d_r, the depth of the tension area given by the Table below for the neutral line, when $C \div T = q$, for the material, and equal to or less than $\frac{2}{3}d$, the diameter.

The volume of *one* pressure may be computed from Eq. 17,

---

* Senate Ex. Doc. No. 1—47th Congress, 2d Session.

page 16, or from computing and tabulating the *volumes* for all required values of d_c, when the radius is *unity*, which we will represent by fC, and since the volumes of cylindrical wedges are as the squares of their radii, we have for the required load, L,

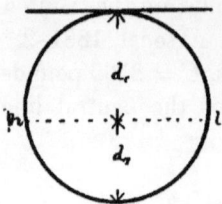

$$L = 2r^2 fC. \qquad (100)$$

Table of positions of the neutral line, d_c, and the computed values of f for the various required values of $q = C \div T$.

q.	d_c.	f.	q.	d_c.	f.
8.0	0.3984 d	0.4877	4.0	0.4951 d	0.6578
7.875	0.4004 d	0.4911	3.866	0.5000 d	0.6666
7.75	0.4022 d	0.4942	3.75	0.5045 d	0.6749
7.625	0.4041 d	0.4975	3.625	0.5095 d	0.6838
7.5	0.4062 d	0.5011	3.5	0.5146 d	0.6947
7.375	0.4087 d	0.5054	3.375	0.5200 d	0.7034
7.25	0.4112 d	0.5097	3.25	0.5256 d	0.7137
7.125	0.4137 d	0.5139	3.125	0.5314 d	0.7239
7.0	0.4161 d	0.5180	3.0	0.5375 d	0.7354
6.875	0.4186 d	0.5221	2.875	0.5438 d	0.7469
6.75	0.4209 d	0.5265	2.75	0.5505 d	0.7691
6.625	0.4236 d	0.5311	2.625	0.5574 d	0.7723
6.5	0.4262 d	0.5355	2.5	0.5646 d	0.7851
6.375	0.4289 d	0.5403	2.375	0.5724 d	0.7993
6.25	0.4316 d	0.5446	2.25	0.5804 d	0.8143
6.125	0.4344 d	0.5496	2 125	0.5890 d	0.8302
6.0	0.4373 d	0.5549	2.0	0 5980 d	0.8469
5.875	0.4402 d	0.5599	1.875	0.6076 d	0.8641
5.75	0.4432 d	0.5652	1.75	0.6178 d	0.8838
5.625	0.4463 d	0.5606	1.625	0.6277 d	0.9042
5.5	0.4494 d	0.5761	1.5	0.6404 d	0.9260
5.375	0.4526 d	0.5819	1.375	0.6525 d	0.9496
5.25	0.4559 d	0.5877	1.25	0.6666 d	0.9747
5.125	0.4594 d	0.5937	1.125	0.6816 d	1.0039
5.0	0.4629 d	0.6000	1.0	0.6976 d	1.0331
4.875	0.4665 d	0.6066	0.875	0.7162 d	1.0661
4.75	0.4702 d	0.6132	0.75	0.7357 d	1.1042
4.625	0.4740 d	0.6199	0.625	0.7583 d	1.1466
4.5	0.4780 d	0.6270	0.5	0.7843 d	1.1946
4.375	0.4821 d	0.6343	0.375	0.8149 d	1.2511
4.25	0.4863 d	0.6419	0.25	0 d	
4.125	0.4906 d	0.6496			

EXAMPLE 48.—Required the breaking load of a Cylindrical Cast-Iron Column, when

The diameter $d =$ $1''.129$, $C = 96280$ lbs. computed in Ex. 3,
" length... $l = 10''.0$, $T = 29400$ " by test,
" deflection $\delta = $ $0''.4$, $q = 3.271$.

The value of f from the Table corresponding to this value of q is $f = 0.715$, then from Eq. 100 we have

$$L = 2 \times (0.5645)^2\, 0.715 \times 96280 = 43884 \text{ pounds.}$$

This column was cast from the iron described in Example 14, page 82, "Cracks developed on the tension side when the deflection reached $0''.4$, the load sustained being 40000 pounds." Four other columns of the same iron were broken, with deflections varying from $0''.38$ to $0''.43$, but the loads sustained were not given. The theoretical determination of the position of the neutral line and the practical is almost identical, the theoretical being $d_\tau = 0.538$, and the practical $d_\tau = 0''.40$.

HOLLOW CIRCULAR PILLARS.

For this Case the deflection, δ, must be equal to or greater than d_τ, the depth of the tension area given by the following Table for the *neutral line*, when $q = C \div T$ for the material, and equal to or less than $\frac{1}{4}d$, the diameter of the outer surface of the pillar.

The volume of one pressure wedge may be computed from Eq. 20, page 20, or from computing and tabulating the volumes for all required values of d_c, when the radius is unity, which we will represent by tf_0C, and since they are as their radii we have for L the load, t being the thickness of the metal—

$$L = 2rtf_0C. \tag{101}$$

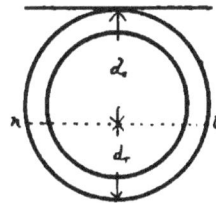

Table of positions of the neutral line, d_c, and the computed values of f_0 for the various required values of $q = C \div T$.

q.	d_c.	f_0.	q.	d_c.	f_0.
8.3195	0.5000 d	2.0000	4.5	0 6520 d	2.3377
8.0	0 5093 d	2.0214	4.375	0.6590 d	2.3528
7.875	0.5131 d	2.0299	4.25	0 6662 d	2.3684
7.75	0.5170 d	2.0389	4.125	0.6735 d	2.3845
7.625	0.5209 d	2.0481	4.0	0.6811 d	2.4010
7.5	0.5249 d	2.0566	3.875	0.6889 d	2.4180
7.375	0.5290 d	2.0657	3.75	0.6970 d	2.4355
7.25	0 5331 d	2.0752	3.625	0.7052 d	2.4533
7.125	0.5374 d	2.0847	3.5	0.7137 d	2.4717
7.0	0.5418 d	2.0945	3.375	0.7223 d	2.4906
6.875	0.5462 d	2.1044	3.25	0.7313 d	2.5101
6.75	0.5507 d	2.1144	3 125	0.7405 d	2.5301
6.625	0.5553 d	2.1248	3.0	0.7500 d	2.5509
6.5	0.5600 d	2.1351	2.875	0.7597 d	2.5722
6.375	0.5647 d	2.1457	2.75	0.7697 d	2.5941
6.25	0.5697 d	2.1568	2 625	0.7800 d	2.6167
6.125	0.5748 d	2.1680	2.5	0.7906 d	2.6400
6.0	0.5800 d	2.1790	2.375	0.8014 d	2.6638
5.875	0.5852 d	2.1910	2.25	0.8122 d	2.6877
5.75	0.5907 d	2.2035	2.125	0.8241 d	2.7143
5.625	0.5962 d	2.2151	2.0	0.8358 d	2.7404
5.5	0.6018 d	2.2277	1.875	0.8478 d	2.7673
5.375	0.6076 d	2.2104	1.75	0.8600 d	2.7950
5.25	0.6134 d	2.2530	1.625	0.8724 d	2.8234
5.125	0.6193 d	2.2663	1.5	0 8851 d	2.8526
5.0	0.6255 d	2.2798	1.375	0.8979 d	2.8823
4.875	0.6314 d	2.2933	1.25	0.9107 d	2.9124
4.75	0.6384 d	2.3079	1.125	0.9233 d	2.9425
4.625	0.6451 d	2.3226	1.0	0.9360 d	2.9732

EXAMPLE 49.—Required the Greatest Strength of a Phœnix Column, when

The outer diameter ....$d = 8''.00$, $q = 1$,
  "   thickness of metal, $t = 0''.35$, $f_0 = 2.9732$ for $q = 1$,
  "   deflection........$\delta = 0''.535$, $l = 7'.0$.

The crushing value of $C = 60000$ is assumed, then from Eq. 101,

$$L = 2 \times 4.0 \times 0.35 \times 2.9732 \times 60000 = 499497 \text{ pounds.}$$

By the United States Government Watertown Arsenal tests the greatest strength was 468000 pounds.*

117. Case V.—PILLARS THAT CROSS-BREAK FROM COMPRESSION AND CANTILEVERAGE.

The *shortest* pillar of this class, for a given section and material, is that length that just deflects sufficiently to allow the resultant of the applied load to pass to the right of the fulcrum f (Fig. 34). This for a given material varies with the section, and the class includes all pillars of longer lengths. The amount of direct pressure along the line fn is equal to the *sum* of the two equal pressure wedges described in Case IV. for the given cross-section, and is the same for all deflections, but to obtain the breaking load of the pillar this *sum* must be diminished by the tensile and compressive strain that arises from the load's action as a cantilever.

C' = the greatest intensity of the compressive strain arising from bending action of the load.

RECTANGULAR PILLARS.

$\delta - \tfrac{1}{3}d$ = the lever-arm of the load, L.

Then the moment of the applied load will be, from Eq. 23, p. 38,

$$L\left(\delta - \tfrac{1}{3}d\right) = \frac{bd_c^2 C'}{3},$$

$$\therefore\, C' = \frac{L\left(3\delta - d\right)}{bd_c^2}.$$

Then we will have for the applied load, L,

$$L = bd_c C - bd_c C',$$

---

* Ex. Doc. No. 23, House of Representatives, 46th Congress, 2d Session, p. 278.

from which, by substituting the above value of C', we obtain

$$L = \frac{bd_c C}{1 + \frac{3\delta - d}{d_c}}.$$ (102)

When 3δ becomes equal to or *less* than d, the depth, the formula becomes that for rectangular pillars of Case IV., the cantilever effect of the load having disappeared.

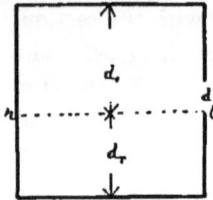

As before stated, the formulas deduced for the strength of columns that fail by cross-breaking with cantileverage are identical in form with those given by Mr. Lewis Gordon, but, unlike his, they give exact values and show when the formula does not apply. This uncertainty as to the length of pillars to which Gordon's formulas did not apply has been the chief objection to their use in practice.

Adapting Gordon's formula for the strength of rectangular pillars to our notation, we have

$$L = \frac{AC}{1 + c\,\frac{l^2}{d^2}}.$$

From making A the full area of the column, the factor C could never be the crushing strength of the material, but is always something less than it in value. The second member of the denominator makes the formula true for all lengths of pillars, which experiment does not confirm, and is of the same general form as our formula for the deflection given by Eq. 93.

No successful effort has ever been made to so modify Gordon's formula that the computed and experimental strength of wooden columns would be the same in value; and for the reason that in all published results of experiments on wooden pillars, except for a few on yellow pine made at the Watertown Arsenal, they all failed without cantileverage of the applied load, and the deflection did not enter as a factor to decrease the strength as contemplated by Gordon.

In the tests of white and yellow pine columns made at the Watertown Arsenal, and described in Examples 46 and 47, the strength was not decreased · from cantileverage. For any given rectangular section of these pillars, the strength is the same for all lengths from twenty-seven to forty-five times the least diameter or least side of the rectangle, which includes the lengths of all yellow and white pine pillars that are used in structures.

EXAMPLE 50.—Required the breaking load of a Rectangular Wrought-Iron Column, tested with two $1'\frac{1}{2}$ pin ends, when

The diameter.....$d = 3''.0$, $C = 50000$ pounds assumed.
  " breadth......$b = 3''.0$, $T = 50000$ pounds mean tests.
  " deflection....$\delta = 1''.61$, $q = 1$.

The position of the neutral line from the Table, page 92, is $d_c = 0.78d$; from Eq. 102 we have

$$L = \frac{3 \times 2.34 \times 50000}{1 + \dfrac{3 \times 1.61 - 3.0}{2.34}} = \frac{351000}{1 + 0.783} = 197000 \text{ pounds.}$$

This example and those in the following Table are from the series of tests described in Example 44, page 128, as given in " Senate Ex. Doc. No. 5—48th Congress, 1st Session," pp. 68 to 102.

Dimensions 3″ × 3″.

$l \div d.$	Length in Inches.	DEFLECTION IN INCHES.		LOAD IN POUNDS.	
		From.	To.	Computed.	Experimental.
28	84	0.5	......	....	216000
30	90	0.5	......	... .	218975
32	96	0.5	1.61	197000	219000
34	102	0.3	1.61	197000	208500
36	108	0.3	1.80	173760	190500
38	114	0.5	1.80	173760	181125
40	120	0.6	1.9	163000	181750
42	126	0.5	1.67	188800	169700
44	132	0.4	2.30	131700	178950
46	138	0.4	1.98	155600	157325
48	144	0.45	2.16	141100	155790
50	150	0.45	2.30	131700	153065
52	156	0.35	2.50	120088	155000
54	162	0.30	2.50	120088	147750
56	168	0.30	2.80	106000	149875
58	174	0.35	2.50	120088	128775
60	180	0.30	2.50	120088	126875

When under the greatest load sustained by the above columns, they "suddenly sprung" to a *cross-breaking* deflection with *cantileverage*. The "Deflection from" in the Table is that at which the "sudden spring" began, and "Deflection to" is that to which it *sprung;* the deflection in the Table is the mean of those of two tests, in most cases.

The "Experimental" load is the mean of two tests, in each case, and is that sustained by the column at the beginning of the "sudden spring." From the "Computed" loads it will be seen that the *momentum* of the "sudden spring" caused the observed deflections to exceed the *true* deflection in many examples of this series of tests.

CIRCULAR PILLARS.

$\delta - \frac{3}{8}d =$ the lever-arm of the load, L,

$f =$ the factor in Table, page 134,

$f_c =$ the   "   "   "   pages 80, 100 and 112.

From Eq. 76, page 57, the moment of the applied load, L, becomes

$$L\left(\delta - \tfrac{3}{8}d\right) = r^3 f_c C',$$

$$\therefore C' = \frac{L\left(8\delta - 3d\right)}{r^3 f_c},$$

and for the load from the pressure wedges given by Eq. 100 we have

$$L = 2r^3 fC - 2r^3 fC'.$$

Substituting for C' its value given above, we have

$$L = \frac{2r^3 fC}{1 + \dfrac{2f(8\delta - 3d)}{r f_c}}. \tag{103}$$

When 8δ becomes equal to or less than $3d$ the load must be computed from the formula for Circular Pillars, Case IV., as in this position of the pillar there will be no cantilever effect of the applied load to be deducted from the sum of the two pressure wedges.

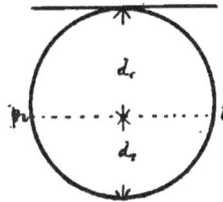

HOLLOW CIRCULAR PILLARS.

$d =$ the outer diameter in inches,
$r =$ the " radius " "
$(\delta - \tfrac{1}{4}d) =$ the lever-arm of the load, L,
$f_0 =$ the factor given in Table, page 136,
$f_c =$ the " " " pages 85 and 104.
$t =$ the thickness of the metal ring in inches.

From Eq. 91, page 62, the moment of the applied load becomes

$$L\left(\delta - \tfrac{1}{4}d\right) = r^2 t f_c C',$$

$$\therefore C' = L\frac{(4\delta - d)}{4r^2 t f_c}.$$

For the load, L, we have

$$L = 2rtf_0 C - 2rtf_0 C'.$$

Substituting the value of C', we obtain

$$L = \frac{2rtf_0 C}{1 + \dfrac{f_0(4\delta - d)}{2rf_c}}. \qquad (104)$$

When 4δ is equal to or less than the outer diameter, d, the strength of the column must be computed from Eq. 101, Case IV., as there will then be no cantilever effect to deduct.

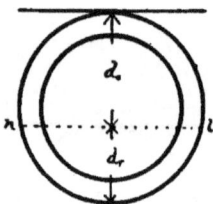

EXAMPLE 51.—Required the Breaking Strength of a Phœnix Column tested with flat ends; $C = T = 60000$ is the assumed strength of the iron.

Outer diameter....$d = 8''.0$, $q = C + T = 1$,
Thickness of metal, $t = 0''.35$, $f_0 = 2.9732$ from Table, page 136, for $q = 1$,
Deflection.........$\delta = 2''.47$, $f_c = 2.6806$ "   "   " 104, " $q = 1$.

From Eq. 104 we have

$$L = \frac{2 \times 4 \times 0.35 \times 2.9732 \times 60000}{1 + \dfrac{2.9732(4 \times 2.47 - 8)}{2 \times 4 \times 2.6806}} = \frac{499497}{1+0.26} = 396426.$$

This example is taken from the series of experiments* described in Example 49, page 136; the tested breaking load was 416000 pounds; the length was 28 feet.

---

* Report of the United States Board appointed to test iron and steel, Ex. Doc. No. 23, House of Representatives, 46th Congress, 2d Session, page 270.

ANGLE-IRON, BOX, CHANNEL, EYE-BEAM AND TEE PILLARS.

In pillars of the above sections, the distance of the neutral line from that side of the pillar that will most likely be its concave side, when broken, must be computed by the rules heretofore given for it in beams of the section of the pillar. The *sum* of the two pressure wedges will then be the product of the compressed area, A_c, by the crushing strength of the material.

The distance, g, of the centre of gravity of the applied load wedge from the point k (Fig. 34), must be determined, which is obtained by computing RF, the *moment* of the load wedge with respect to the axis, k, in which F is the greatest intensity of the load at k, it being zero at m, and R a factor that depends for its value upon the section; for any given dimensions it reduces to a numerical quantity without assigning any value to F; and dividing this moment, RF, by the volume of the load wedge, $AF \div 2$, in which A is the area of the section, we have

$$g = RF \div \frac{AF}{2} = \frac{2R}{A},$$

and

$$L(\delta - g) = R_c C' \therefore C' = \frac{L(\delta - g)}{R_c}.$$

Then we have

$$L = A_c C - A_c C'.$$

Substituting the above value of C', we deduce

$$L = \frac{A_c C}{1 + \dfrac{A_c(\delta - g)}{R_c}}. \tag{105}$$

R_c is a factor of the *moment* of resistance that the section of the pillar offers to the cantilever bending of the load; for a given section it becomes a numerical quantity without assigning any value to C' when computed by the rules given for the moment of resistance of the section when strained in a beam.

When δ becomes equal to or less than g in value the cantilever strain ceases to exist, and the pillar belongs to Case IV.,

$$\therefore L = A_c C. \tag{106}$$

When δ becomes less than d_r, the depth of the extended area required for rupture, the pillar belongs to Case III.

$$\therefore L = \left[C + q \frac{\delta T}{d_r} \right] \frac{A_c}{2}, \tag{107}$$

in which d_r gives the position of the neutral line of rupture, and q the ratio of the compressive strain that will be required to hold in equilibrium the tensile strain developed by the bending of the pillar as a cantilever.

Equation 105 is the general formula for the strength of pillars of all lengths, sections and material of which the formulas heretofore deduced in this chapter are only the forms it will assume for special cases. The denominator of the second member of the formula must never be less than *unity*.

CHAPTER VIII.

118. General Statement. In roof-trusses, cranes, derricks, platforms supported by cantilevers, trussed beams and other structures, there is used a class of pieces of material that, from the manner in which they are loaded, do not belong exclusively to either horizontal beams or columns, but partake of the nature of both, in the manner in which they support the load to which they are subjected.

The theory of the *transverse* strength of these Combined Beams and Columns gives the solution of the general problem of the transverse strength of all beams, without regard to the special angle that the axis of the beam makes with the line of direction of the loading and supporting forces. Horizontal beams acting under vertical loads and columns are only special cases of the general problem, in which certain factors, that cause the strength of the same piece of material to vary with its angle of inclination to the horizon, disappear, from the general rule for these cases, by becoming zero in value. But on account of their great importance, we have, in the preceding chapters, deduced separately the principles and rules from which the strength of these special cases may be computed.

. Should one end of a horizontal beam, such as *be* (Fig. 44), be fixed in a vertical wall, and a load be *attached* to its free end, not suspended, as in the figure, none of this load will directly compress or *rest* upon the cross-section of the beam, *be*. Now let the wall be revolved around the point *F*, to a horizontal position, thus bringing the beam, *be*, to the vertical, then the entire load will rest upon or directly compress the cross-section of the beam, now converted into a pillar; the load has thus been gradually converted from a non-compressing to a compressing load with its full weight. The bending moment of the *attached* load is greatest when the beam is in the horizontal position, and it gradually diminishes as its lever-arm, *s*, becomes less in value with the revolution of the wall, and becomes zero in value when the beam occupies the vertical position. On the other hand, should the wall be revolved to the horizontal position around the point *P*, the bending moment will gradually decrease and become zero, while the tensile strain will increase from zero to that of the full weight of the load, when the beam becomes vertical.

From the above illustration the origin of the special cases of *horizontal* beams and *columns* is apparent, and the reason for the special rules for their strength. A similar illustration could be deduced from a beam supported at both ends, by conceiving it to occupy all positions from the horizontal to the vertical.

In order to deduce the relation that exists between the *applied load* and the *resistance* of the combined beam and column at the instant of *rupture*, it is *assumed*, in the analysis, that they only *deflect* enough to admit of their *cross-breaking* as a column, without *cantileverage* of the applied load, as in Class IV., Chapter VII. In very long beams, however, the compression resulting from pressure applied to its ends will act with the leverage explained in Case V., page 137, and the strength must be computed from the formulas there given, but the compression resulting from the transverse bending of the load will be the same in each case, and will compress the beam without cantileverage.

119. Notation. In addition to the notation heretofore used and defined in Art. 35, page 37, the following will be used in this Chapter :

$C = C' + C'' =$ the greatest intensity of the compressive strain in pounds per square inch,

$C' =$ the greatest intensity of the compressive strain in pounds per square inch, arising from the bending component of the load,

$C'' =$ the greatest intensity of the compressive strain in pounds per square inch, arising from the compressing component of the applied load,

$L_b =$ the bending component of the applied load in pounds,

$L_c =$ the compressing component of the applied load in pounds,

$l =$ the unsupported length of the beam in inches,

$s =$ the span, the horizontal distance between the supports in inches,

$h =$ the difference between the heights of the ends of the beam in inches,

$a =$ the angle that the beam makes with the horizon.

In many of the different methods of loading and supporting beams given in this Chapter, the greatest resulting compressive strain at any section is the *sum* of two or more distinct pressures, which will be represented by the letter C, with corresponding accents, such as C', C'', C''', C'''', etc.

Inclined Beams.

120. General Conditions. The effect produced by a load when applied to this class of beams will manifest itself in two *distinctly* different ways; each separate effect must be computed, and their sum will be the total effect produced by the load upon the beam. This is accomplished by decomposing the applied load into *two components*, by the well-known theorem of the *parallelogram* of forces. *One component, L_b,* must be at right angles to the axis of the beam, and the *other*, L_c, parallel with it; the first component will bend the piece as a *beam*, while the second will compress it as a *column*.

The parallelogram of forces for inclined beams is a rectangle, and the relation between the components and the load is obtained from that of the three sides in a right-angle triangle, in which

$$cos.\ a = \frac{s}{l} \qquad sin.\ a = \frac{h}{l} \qquad tan.\ a = \frac{h}{s}.$$

121. Inclined Beam Fixed and Supported at One End. This method of "fixing" and loading beams is illustrated by two different positions of the beam, be, in Fig. 44. Two Cases will be considered.

Case I.— *When the load is applied at the free end of the beam.*

Let the right-angle triangle, *eot* (Fig. 44), represent the *parallelogram* of forces, · in which *et* represents the applied

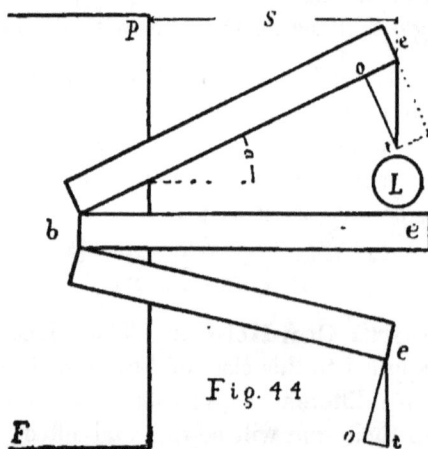

Fig. 44

load, drawn to any given scale, then ot will be the *bending* and oe the *compressing* component of the applied load, L, from which

$$ot = et\ cos.\ a \text{ and } oe = et\ sin.\ a,$$

$$\therefore L_{\text{n}} = \frac{Ls}{l}, \tag{108}$$

and

$$L_c = \frac{Lh}{l}. \tag{109}$$

The bending moment of the component, L_B, its lever-arm being l, will be

$$Bending\,Moment,\, L_B = \frac{Ls}{l} \times l = Ls. \tag{110}$$

Having decomposed the applied load, L, into two components, one perpendicular and the other parallel to the inclined beam, we can now ascertain the effect that will be produced by the original load upon it, by a combination of the methods used to compute the effect produced upon horizontal beams and columns by vertical applied loads.

RECTANGULAR BEAMS.

The *transverse strength* of a rectangular beam loaded and fixed as in this Case will now be deduced from the foregoing formulas.

The bending moment from Eq. 110 must be made equal to the moment of resistance from Eq. 23, page 38, in which $C = C'$.

$$\therefore Ls = \frac{bd_c^2 C'}{3}, \tag{111}$$

$$\therefore \frac{3Ls}{d_c} = bd_c C'. \tag{112}$$

The compressing component, L_c, from Eq. 109, must be resisted by an equal compression produced in the section at the face of the wall, as given by Eq. 99, page 131, in which

$$C = C'' \text{ and } L = \frac{Lh}{l},$$

$$\frac{Lh}{l} = bd_c C''. \tag{113}$$

Adding Eqs. 112 and 113 we have, by making $C' + C'' = C$, the crushing strength of the material,

$$L\left(\frac{3s}{d_c} + \frac{h}{l}\right) = bd_cC, \tag{114}$$

from which

$$L = \frac{lbd_c^2C}{3ls + d_ch}. \tag{115}$$

When the beam is horizontal, $h = 0$ and $l = s$, the formula reduces under this hypothesis to that given in Eq. 27, page 38, in which $m = 1$. In a vertical beam, $s = 0$ and $l = h$, the formula reduces to that given for the strength of rectangular columns, Eq. 99.

To Design a Rectangular Beam.

The length l and height h will be controlled by the position in which the beam is to be used. A convenient depth, d, must then be assumed, and the value of d_c computed from Eq. 26, page 38 ; then from Eq. 114 we obtain

$$b = \frac{(3ls + d_ch)\,L}{ld_c^2C}, \tag{116}$$

from which the required breadth, b, of a beam that will break with a given load, L, will be obtained by giving to C the value of the crushing strength of the material of which the beam is to be constructed.

Case II.—When the Load is uniformly distributed over the unsupported length of the beam.

The *resultant* of the applied load will pass through the middle of the length of the beam and the triangle, *eot* (Fig. 44), will, as in Case I., give the relation between the load and its two components; hence

$$L_s = \frac{Ls}{l}, \tag{117}$$

$$L_c = \frac{Lh}{l}. \tag{118}$$

The bending moment of the component, L_s, will be,

$$Bending\ Moment = \frac{Ls}{l} \times \frac{l}{2} = \frac{Ls}{2}, \qquad (119)$$

its lever-arm being $l \div 2$.

RECTANGULAR BEAMS.

The *transverse* strength of a rectangular beam loaded and fixed as in this Case may be obtained by the same process heretofore used in Case I.,

$$\therefore L = \frac{2lbd_c{}^2C}{3ls + 2d_ch}. \qquad (120)$$

To Design a Rectangular Beam.

Assume a depth, d, and compute d_c from Eq. 26, page 38, then from Eq. 120 we have

$$b = \frac{(3ls + 2d_ch)\,L}{2\,lbd_c{}^2C}. \qquad (121)$$

122. Inclined Beam, supported at one end and STAYED or held in position at the other without vertical support. This class of Inclined Beams is illustrated in Figs. 45, 46 and 47. Five different cases of loading will be considered.

Case I.—WHEN THE INCLINED BEAM IS LOADED AT ITS STAYED END.

The effect produced by the load, L, upon the beam, *be* (Fig. 45), is simply to compress it as a column, the load being held in equilibrium by a *pull* along the tie *ec*, and a thrust along the inclined beam.

In the right-angle triangle, *toe*, let *et* represent the load

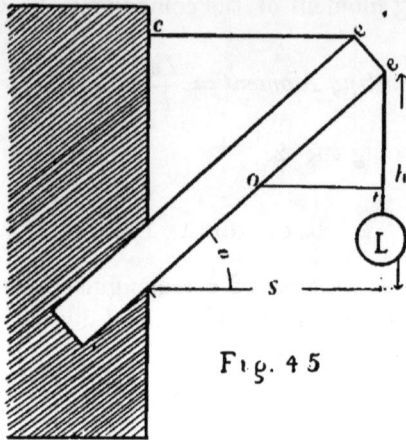

Fig. 45

drawn to any given scale, then *to* gives the pull and *oe* the thrust.

$$ot = \frac{et}{tan.\ a}, \quad oe = \frac{te}{sin.\ a},$$

$$\text{The } pull = \frac{Ls}{h}, \tag{122}$$

and

$$\text{The } thrust\ L_c = \frac{Ll}{h}, \tag{123}$$

from which the compressing effect of the applied load, L, can be ascertained.

RECTANGULAR BEAMS.

The direct compression given by Eq. 123 must be made equal to the resistance offered by the section of the beam as a column, then from Eq. 99, page 131,

$$\frac{Ll}{h} = bd_cC,$$

$$\therefore L = \frac{hbd_cC}{l}. \tag{124}$$

Case II.—When the Inclined Beam is loaded at its Middle.

Let bc (Fig. 46) represent the beam loaded at its middle, t, with the load, L, then in the right-angle triangle, toe, et rep-

Fig. 46.

resents the load drawn to any scale, ot, perpendicular to the beam, bc, the *bending component*, L_B, and ot parallel to bc, the directly *compressing component*, L_c.

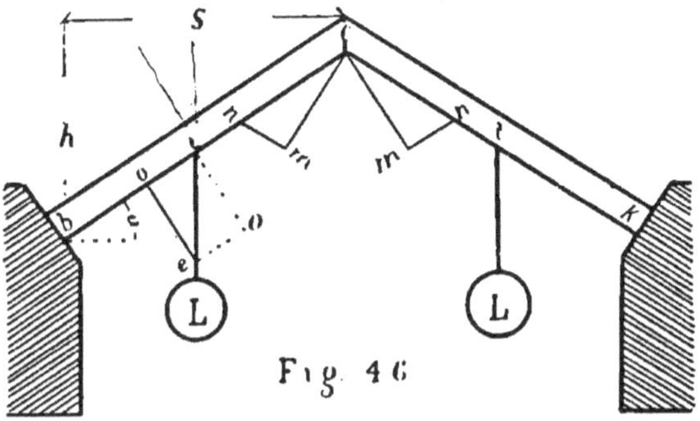

$$L_B = \frac{Ls}{l}, \tag{125}$$

$$L_c = \frac{Lh}{l}. \tag{126}$$

The *bending moment* produced by the component, L_B, is identical with that produced by conceiving the beam to be "fixed" at its middle, t, and loaded at the free end with $L_B \div 2$,

$$\therefore Bending\ Moment,\ L_B = \frac{Ls}{2l} \times \frac{l}{2} = \frac{Ls}{4}. \tag{127}$$

In order that the loading and supporting forces shall be in *equilibrium, one half* of the *bending component*, L_B, must be *supported* at each end of the beam b and c. The half of L_B at c cannot be *directly supported* at that point, but must be *carried* to the *ground* by some means. The method of *staying* the end of the beam, *bc*, represented in the figure is that used in roof-trusses; hence the load, $L_s \div 2$ at *c*, must be decomposed into two components, one in the direction of each beam or rafter; if the angle, *bck*, is a right angle the entire load, $L_s \div 2$, compresses the rafter, *ck*, as a pillar, thus reaching the support, *k*.

In roof-trusses the inclined beams, *bc* and *ck*, are usually loaded in the same manner, and *each* beam will, therefore, carry to *c* a component, $L_s \div 2$, of its load for support, which will be *equivalent* to the component, $L_s \div 2$, compressing each rafter to which it is applied, but much increased in amount from the manner in which it is converted from a load that is perpendicular to the beam into a compressing load that is parallel to its axis.

In the two *equal* triangles, *cmn*, let *cm* in each represent the load, $L_s \div 2$, *cn* the equal components that the rafters *exchange* with each other, and *mn* the equal components that the loads, $L_s \div 2$, produce that strain the rafters to which they are applied. When the angle, *bck*, is greater than 90° the sum of the components $(mn + cn)$ will be greater than $L_s \div 2$, when *bck* = 90°, $(mn + cn) = L_s \div 2$, *mn* becoming zero. When *bck* is less than 90° the component, *mn*, is a *tensile* strain; the total compression $(cm - mn)$ will then be less than $L_s \div 2$.

In roof-trusses the angle, *bck*, is generally greater than 90°, and the *total* compression upon each rafter resulting from the component of the load that it carries to *c* is $(mn + cn)$.

$$mn = mc \,.\, tan. \,(90° - 2a),$$

$$cn = \frac{mc}{cos. \,(90° - 2a)}.$$

Adding and substituting $mc = \dfrac{Ls}{2l}$, we have

$$. (mn + cn) = \frac{Lcs}{2l} = \frac{Ls}{2l}\left(tan.(90° - 2a) + \frac{1}{cos.(90°-2a)}\right). \quad (128)$$

The factor, c, in this equation, for convenience is

$$c = tan.(90° - 2a) + \frac{1}{cos.(90° - 2a)}. \quad (129)$$

When the angle $a = 45°$, $tan.(90° - 2a) = 0$, $\frac{1}{cos.(90° - 2a)} = 1$,

and the amount of compression will become $\frac{L_\text{B}}{2} = \frac{Ls}{2l}$, the

angle, bck, being a right-angle.

The total compression produced by the single load, L, at the middle of each rafter will be the sum of the *three* distinct parts, represented by Eqs. 126, 127 and 128.

Rectangular Beams.

The amount of *direct* compression from Eqs. 126 and 128 must be equal to the resistance of the beam as a column, from Eq. 99, page 131, in which $C = C'$ and $C = C'''$ respectively.

$$\therefore \frac{Lh}{l} = bd_cC', \quad (130)$$

and

$$\frac{Lcs}{2l} = bd_cC'''. \quad (131)$$

The *bending moment* from Eq. 127 must be equal to the *moment* of *resistance* of the section as a beam from Eq. 23, page 38, in which $C = C''$.

$$\therefore \frac{Ls}{4} = \frac{bd_c^2C}{3}, \quad (132)$$

$$\therefore \frac{3Ls}{4d_c} = bd_cC''. \quad (133)$$

Adding these separate components, as given by Eqs. 130,

131 and 133, and making $C' + C'' + C''' = C$, the greatest compressive strength of the material, we have

$$L\left(\frac{3s}{4d_c} + \frac{h}{l} + \frac{cs}{2l}\right) = bd_cC, \qquad (134)$$

$$\therefore L = \frac{4\ l\ bd_c^2 C}{3sl + (4h + 2cs)\ d_c}, \qquad (135)$$

from which the required breaking load may be computed.

Case III.—When the Load is uniformly distributed over the length of the Inclined Beam.

The bending moment and the component L_c will be identical with those produced by conceiving *one half* of the total load, L, to be concentrated at its middle section, t (Fig. 46); then, from Eqs. 125, 126 and 127, Case II., by making $L = L \div 2$, we have

$$L_{\text{B}} = \frac{Ls}{2l}, \qquad (136)$$

$$L_c = \frac{Lh}{2l}, \qquad (137)$$

$$\textit{Bending Moment, } L_{\text{B}} = \frac{Ls}{8}. \qquad (138)$$

The bending moment is produced by a component of the load, L, that is perpendicular to the length of the beam, and uniformly distributed over its length, bc. In order that equilibrium shall exist, one half of this uniformly distributed component, $\frac{Ls}{l}$, must be *supported* at each end of the beam, and as it cannot be *directly supported* at the *stayed* end, c, it must be carried to the supports b and k, as described in Case II., Eq. 128.

The method of analysis of the strains in a simple roof-truss, given in this and the preceding Case, differs from that usually pursued by writers. One half of the load on each rafter is usually considered to be supported by b and k, and the other half of each rafter load to be supported at c, by reacting against each other. The transverse strength is then computed as if the rafter were a horizontal beam, which is equivalent to saying, that for a given material, section and span of roof, as in Fig. 46, the transverse strength would be the same for the infinite number of roof-trusses that could be constructed between these points of support by varying the *pitch* or angle that the rafter makes with the horizon, which, of course, is not the case.

Rectangular Beams.

From Eqs. 128, 137 and 138, we obtain, by a process similar to that used in Case III.,

$$L\left(\frac{h}{2l} + \frac{cs}{2l} + \frac{3s}{8d_c}\right) = bd_c C, \qquad (139)$$

$$\therefore L = \frac{8lbd_c^2 C}{3sl + (4h + 2cs)d_c}. \qquad (140)$$

Example 52.—Required the uniformly distributed breaking load of a White Pine Rafter, as in Fig. 46, when

The span......$s = 15'. 0$, $C = 5000$ pounds mean of tests,
" length.....$l = 16'. 8$, $T = 10000$ "   "   "   "
" rise.......$h = 7'. 5$, $g = 0.5$,
" depth.....$d = 9''.0$, $d_c = 0.868d$, from Table, page 108,
" breadth...$b = 5''.0$, $a = 26°.34'$.

From Eq. 129, $c = 2$, then, from Eq. 140,

$$L = \frac{8 \times 201.6 \times (7.812)^2 \times 5000}{3 \times 180 \times 201.6 + (4 \times 90 + 2 \times 2 \times 18).868 \times 9} = 20977 \text{ lbs.}$$

This Example is taken from Trautwine's "Engineer's Pocket-Book," and was designed to sustain an uniformly distributed load of 8000 pounds, with a factor of safety of three.

Case IV.—When the Inclined Beam is Loaded at its middle, and with an additional Load at its stayed end.

The bending moment and the compression produced by the load, L, applied at t, the middle of the beam, bc (Fig. 47), will

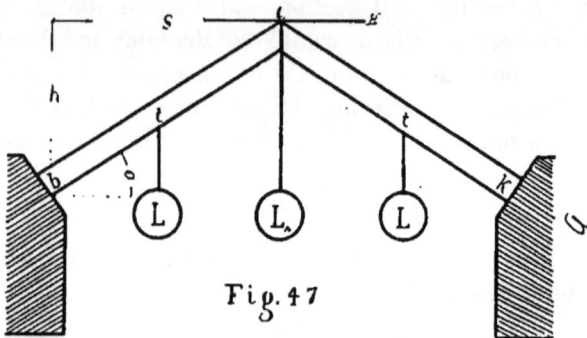

Fig. 47

be obtained from Eqs. 126, 127, and 128 of Case II., the two conditions of loading being the same.

One half of the additional load, L_A, suspended from the *stayed* end, c, is supported by each support, b and k; the compression produced upon the inclined beam as a column has been deduced in Case I., Eq. 123; the horizontal components, H and F, acting in opposite directions, neutralize each other.

From Eqs. 122 and 123, page 152, we have, by making $L = L_A \div 2$,

$$\textit{Horizontal component, } H = \frac{L_A s}{2h}, \qquad (141)$$

$$\textit{The Thrust, } L_c = \frac{L_A l}{2h}. \qquad (142)$$

Rectangular Beams.

The amount of *direct* compression from Eqs. 126, 128 and 142 must be made equal to the *resistance* of the beam as a column, as given in Eq. 99,

$$\frac{Lh}{l} = bd_c C', \qquad (143)$$

$$\frac{Lcs}{2l} = bd_c C''', \qquad (144)$$

$$\frac{L_\Lambda l}{2h} = bd_c C''''. \qquad (145)$$

The *bending moment* from Eq. 127 must be equal to the *moment* of *resistance* of the section of a beam from Eq. 23, page 38, in which $C = C''$,

$$\frac{Ls}{4} = \frac{bd_c^2 C''}{3}, \qquad (146)$$

$$\therefore \frac{3Ls}{4d_c} = bd_c C''. \qquad (147)$$

Adding Eqs. 143, 144, 145 and 147, and making $C' + C'' + C''' + C'''' = C$, the crushing strength of the material, we have

$$L \left(\frac{3s}{4d_c} + \frac{h}{l} + \frac{cs}{2l} \right) + \frac{L_\Lambda l}{2h} = bd_c C, \qquad (148)$$

$$L = \frac{2ld_c (2hbd_c C - L_\Lambda l)}{h (3sl + 2d_c (2h + cs))}, \qquad (149)$$

from which the centre breaking load of the beam may be computed, in addition to that of L_Λ suspended directly from its *stayed* end.

Case V.—When the Load is uniformly distributed over the length of the Inclined Beam, and an additional Load, L_Λ, applied to its stayed end.

This Case is Cases III. and IV. combined, therefore Eqs. 128, 137, 138 and 142 will be the formulas required for computing the transverse strength of the beam.

Rectangular Beams.

The *direct* compression from Eqs. 128, 137 and 142 must be equal to the *resistance* of the beam, as a column, from Eq. 99, page 131.

$$\frac{Lh}{2l} = bd_cC', \tag{150}$$

$$\frac{L_{\wedge}l}{2h} = bdC''', \tag{151}$$

$$\frac{Lcs}{2l} = bd_cC''''. \tag{152}$$

The *bending moment* from Eq. 138 must be equal to the *moment* of *resistance* of the section as a beam from Eq. 23, page 38, in which $C = C''$,

$$\frac{Ls}{8} = \frac{bd_c^2C''}{3}, \tag{153}$$

$$\frac{3Ls}{8d_c} = bd_cC''. \tag{154}$$

Adding Eqs. 150, 151 and 152, we obtain by making $C' + C'' + C''' + C'''' = C.$

$$L\left(\frac{3s}{8d_c} + \frac{h}{2l} + \frac{cs}{2l}\right) + \frac{L_{\wedge}l}{2h} = bd_cC, \tag{155}$$

$$\therefore L = \frac{4ld_c\left(2hbd_cC - L_{\wedge}l\right)}{h\left(3sl + 4d_c\left(h + cs\right)\right)}. \tag{156}$$

123. Inclined Beam Supported vertically at both ends and loaded at its middle.

Let Fig. 49 represent the beam, supports and the load, L. In the triangle, *toe*, let *eo*, to any scale represent the load, L, then the *component*, *ot*, at right angles to the beam, will bend it and *et* parallel to the beam will compress it as a column.

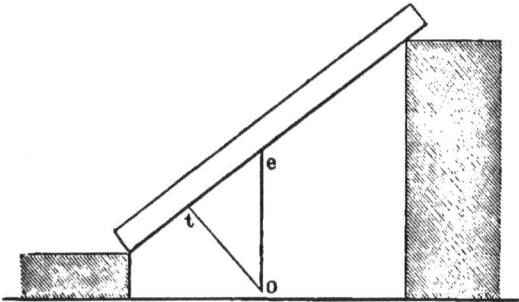

Fig. 49.

$$ot = eo\ cos.\ a \text{ and } ot = eo\ sin.\ a,$$

$$\therefore L_\text{B} = \frac{Ls}{l}, \tag{157}$$

and

$$L_\text{c} = \frac{Lh}{l}. \tag{158}$$

One half of the bending component, L_B, will be supported by the walls under each end of the beam; the bending moment will therefore be

$$Bending\ Moment,\ L_\text{B} = \frac{Ls}{2l} \times \frac{l}{2} = \frac{Ls}{4}. \tag{159}$$

Having decomposed the load into two components, one perpendicular and the other parallel to the axis of the inclined beam, we can compute its effect upon a beam of any section and material.

LOAD SUSTAINED BY THE SUPPORTS.

Each of the walls, supporting the ends of the inclined beam, must resist a thrust of one half of the bending component, L_B, which tends to overturn them. Decomposing this thrust at each support into a vertical and horizontal component, the first will be the proportion of the centre load, L, that is sustained by the higher support. The lower support also sustains an equal vertical component, and in addition a vertical component of the compressing component of the applied load, L.

The horizontal thrust against the higher support will be the horizontal component of one half of L_{a}; that at the lower support will be the sum of the two horizontal components arising from decomposing L_{c} and $L_{\text{a}} \div 2$ at the lower support.

From the above we obtain,

Vertical load on higher support $= \dfrac{Ls}{2l} \cdot \cos. a,$

Vertical load on lower support $= \dfrac{L}{l} \left(\dfrac{s}{2} \cos. a + h \sin. a \right).$

When the beam becomes horizontal the angle $a = o$, $s = l$, $\cos. a = 1$ and the $\sin. a = o$; substituting the values in the above formulas, we find that each wall supports one half of the load, L, as in Case III., page 4.

Rectangular Beams.

The direct compression from Eq. 158 must be equal to the resistance of the beam as a column from Eq. 99, page 131.

$$\frac{Lh}{l} = bd_{\text{c}}C'. \qquad (160)$$

The *bending moment*, from Eq. 159, must be equal to the *moment* of *resistance* of the beam, from Eq. 23, page 38.

$$\frac{Ls}{4} = \frac{bd_{\text{c}}^{2}C''}{3}, \qquad (161)$$

$$\therefore \frac{3Ls}{4d_{\text{c}}} = bd_{\text{c}}C''. \qquad (162)$$

Adding Eqs. 160 and 161 we have, by making $C' + C'' = C$, the compressive strength of the material.

$$L\left(\frac{3s}{4d_c} + \frac{h}{l}\right) = bd_cC, \tag{163}$$

$$\therefore L = \frac{4lbd_c^2C}{3sl + 4hd_c}, \tag{164}$$

from which the centre breaking load of any rectangular inclined beam supported at both ends may be computed.

124. Inclined Beams supported at both ends and the load uniformly distributed over its length.

The bending moment and compression are identical with that produced by conceiving the beam to be "fixed" at its middle and loaded on its free ends with an uniformly distributed load, $L \div 2$, as in Case II., page 150, L being equal to $L \div 2$

$$L_n = \frac{Ls}{2l} \text{ and } L_c = \frac{Lh}{2l}.$$

The bending moment will be,

$$Bending \; Moment, \; L_n = \frac{Ls}{2l} \times \frac{l}{4} = \frac{Ls}{8}.$$

From these components the effect of the uniformly distributed load on an inclined beam of any section and material may be computed as in Art. 123.

LOAD SUSTAINED BY THE SUPPORT.

For an uniformly distributed load, L, the amount of vertical pressure that will be sustained by each wall will be obtained from the formulas before deduced for a concentrated load, L, at the middle of the span, the proportion of the total load sustained by the walls being the same in each case of loading.

The method usually given by writers for determining the load supported by each wall, and the bending moment of the load applied to an inclined beam, is to consider one half of the

total load to be sustained by each wall and the bending moment to be the same as if the beam were horizontal, the span
used in the computation being the horizontal distance, s,
between the supports. This method of analysis is manifestly
incorrect, as it would require that an inclined beam of a given
scantling should have a transverse strength that would neither
be varied by its length nor the angle that it makes with the
horizon, provided its horizontal span remained the same. Or
that an inclined beam that is *infinitely* long, and a horizontal
beam whose span is the distance between the supports of the
inclined beam, will have the same transverse strength, provided the cross-section and material are the same in each beam.

Rectangular Beams.

$$\therefore L_c = \frac{Lh}{2l} = bd_c C', \tag{165}$$

$$\frac{Ls}{8} = \frac{bd_c^2 C''}{3} \quad \therefore \quad \frac{3Ls}{8d_c} = bd_c C''. \tag{166}$$

Adding Eqs. 165 and 166, we have, by making $C' + C'' = C$,

$$L = \frac{8lbd_c^2 C}{3sl + 4d_c h}. \tag{167}$$

Trussed Beams.

125. General Conditions. In trussed beams and
roof-trusses, beams are frequently subjected to both *transverse* and *longitudinal* strains either of compression or tension,
thus acting in the double capacity of a horizontal beam and a
vertical column.

Rectangular Beams.

Let Fig. 48 represent such a rectangular beam loaded
transversely with a *total* load, L, and longitudinally with
either a compressive or tensile load, L_c.

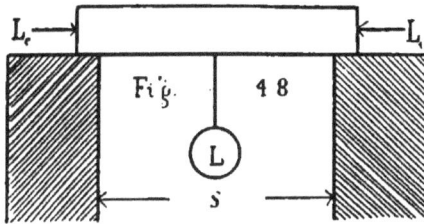

From Eq. 27, page 38, in which $C = C'$, we have

$$L = \frac{mbd_c^2 C'}{3s}, \qquad (168)$$

$$\therefore \frac{3Ls}{md_c} = bd_c C'. \qquad (169)$$

And from Eq. 99, page 131, we have

$$L_c = bd_c C''. \qquad (170)$$

Adding Eqs. 169 and 170, we have, by making $C' + C'' = C$,

$$\frac{3Ls}{md_c} + L_c = bd_c C. \qquad (171)$$

From which either L, L_c, or b may be computed when the other two are known.

$$L = \frac{(bd_c C - L_c)}{3s} md_c, \qquad (172)$$

$$L_c = \frac{mbd_c^2 C - 3Ls}{md_c}, \qquad (173)$$

$$b = \frac{3Ls + md_c L_c}{md_c^2 C}. \qquad (174)$$

HOLLOW CIRCULAR BEAMS.

The relation between the *transverse* load, L, and the resulting *moment* of *resistance* of the beam is given by Eq. 91, page 62, in which $C = C'$.

f_c = the factor from the Table, pages 81 and 104,
f_0 = the   "   "   "   "   page 136.

$$L = mr^2 t f_c C' = \frac{2mr^2 t f_0 f_c C'}{2f_0},\qquad(175)$$

$$\therefore \frac{2f_0 L}{mrf_c} = 2rtf_0 C'.\qquad(176)$$

The compression from the longitudinal strain, L_c, must be equal to the resistance of the beam as a column from Eq. 101, page 135, in which $C = C''$.

$$L_c = 2rtf_0 C''.\qquad(177)$$

Adding Eqs. 176 and 177 we have

$$\frac{2f_0 L}{mrf_c} + L_c = 2rtf_0 C,\qquad(178)$$

$$\therefore L = \frac{mrf_c\,(2rtf_0 C - L_c)}{2f_0}.\qquad(179)$$

The compressing load, L_c, in the foregoing formulas, in practice, generally results from the transverse load, L, its effect being transmitted to the ends of the beam by a vertical post placed under the centre of the beam and connected with the ends by inclined truss-rods. For a single load, L, at the middle it is usually the practice to consider one half of it as producing a tensile strain on each of the inclined truss-rods; this is only strictly correct when the beam is cut into two pieces at its centre. When the beam is continuous it can only strain the vertical post after it begins to deflect; therefore the load that would have caused the untrussed beam to deflect from the horizontal position could bring no strain upon the post.

INDEX.